MW01107579

The Quest for SPEED

Patrick Stephens Limited, a member of the Haynes Publishing Group, has published authoritative, quality books for enthusiasts for more than a quarter of a century. During that time the company has established a reputation as one of the world's leading publishers of books on aviation, maritime, military, model-making, motor cycling, motoring, motor racing, railway and railway modelling subjects. Readers or authors with suggestions for books they would like to see published are invited to write to: The Editorial Director, Patrick Stephens Limited, Sparkford, Nr Yeovil, Somerset, BA22 7JJ.

The Quest for SPEED

Modern racing car design and technology

Alan Henry

First published in 1993

British Library Cataloguing-in-Publication Data:
A catalogue record for this book is available from the British Library.

ISBN 1 85260 437 9

Library of Congress catalog card no. 93-79171

Patrick Stephens Limited is a member of the Haynes Publishing Group PLC, Sparkford, Nr Yeovil, Somerset, BA22 7JJ.

Printed in Great Britain by Butler & Tanner Ltd of London and Frome.

CONTENTS

ACKNOWLEDGEMENTS

The author would like to thank the following for their advice, information and assistance in the preparation of this book:

Bruce Ashmore and Brian Sims (Lola Cars); John Barnard (Ferrari Design & Development); Andy Brown; Brian Hart; Denis Jenkinson; Nigel Roebuck; Charles Bulmer; John Zimmermann (Racer Communications Inc); Gordon Kirby; Dr Gary Savage and Peter Stayner (McLaren International Ltd); Patrick Head, Adrian Newey and Gary Crumpler (Williams Grand Prix Engineering); Dick Scammell and Denise Proctor (Cosworth Engineering); Mario Illien and Line Page (Ilmor Engineering); Fiona Miller (Jaguar Sport); Nick Goozee, Nigel Bennett and Teddy Mayer (Penske Cars); Peter Collins and Peter Wright (Team Lotus); Dr David Hurst (University of Southampton, Department of Aeronautics and Astronautics); Roger Lindsay, Gavin Choyce, Mike Branigan and Tony Dennis (Shell International); Ron McQueeney and Pat Jones (Indianapolis Motor Speedway).

Photography supplied by LAT Photographic; Gary Gold; McLaren International; Benetton Formula; Ilmor Engineering; Cosworth Engineering; Ford Motor Company; and Jaguar.

INTRODUCTION

From the end of the 1970s through to the present day, the technology involved in racing car design and manufacture has accelerated at a quite remarkable rate. This can be seen not only in terms of the complexity of the materials involved in the construction and the sophisticated approach applied to the actual design process, but also in the context of the operational sophistication of the cars in a competitive environment.

This metamorphosis has been brought about for a variety of reasons. Most engineering processes develop within guidelines which are well defined and delineated, but this has not always been the case on the international motor racing scene. The sport has been subjected to outside pressures which have changed its entire complexion, and such changes have manifested themselves in a distinctly different approach to the whole design process.

Consider this analogy. In the early 1950s, the de Havilland Comet airliner catapulted Britain into the very forefront of commercial passenger jet aviation. Most unfortunately, its development was blighted by two major disasters caused by structural failure, as a result of which the American-built Boeing 707 was allowed sufficient development time to catch up and grasp the initiative.

More than 30 years later, the intensive pace of aerospace development had produced such technological marvels as Concorde and the Boeing 747-400. These aircraft, aerodynamically superior, more economical, considerably more reliable whilst at the same time infinitely more complex than their forebears, now represent leading edge aviation technology. They are better and safer than the planes of a generation ago, simply because the application of high technology design processes, superior manufacturing techniques and more sophisticated technology guarantees that they are.

So it is with racing cars. It is obvious that the 1992 Williams-Renault FW14B, which Nigel Mansell drove to win the World Championship, is a vastly superior car to the BRM P57, for example, which Graham Hill drove to the title 30 years earlier. It is more complex, sturdier, more reliable and more sophisticated. Following the analogy, moreover, it is also much, much safer.

Therein lies the crucial difference when one considers the development of racing car technology over the past two decades. Whereas everybody involved in aviation – from Comet to Concorde – obviously regarded passenger safety as so fundamental as to be implicit in everything they did, no such constraining influence fettered the racing car designer. Thirty years ago, racing drivers were involved in a dangerous game, and if they got killed, well, that was a legitimate risk involved in their calling.

A quick trawl through the news columns of motor racing magazines dating from the mid-1960s provides a chilling counterpoint for those schooled solely in the motor racing techniques of the 1980s. Obituary notices tended to be weekly occurrences. There was seldom any technical post mortem, either. A car crashed, a car suffered a component failure, a car caught fire, and the driver died.

That was what motor racing was all about. Deadly risk was part of the calling, and the cars – without safety fuel tanks, without worthwhile rollover bars and without seat harnesses – reflected the thinking of the times. The cars were designed solely with performance considerations in mind. Safety was an incidental aspect. Most people reckoned that the business was as safe or as dangerous as the competing drivers allowed it to be.

Ironically, it would be the presence of television that triggered a long-term sea change in the motor racing design process. One crucial turning point can now be seen to have been in the 1967 Monaco Grand Prix. Lorenzo Bandini, Italy's darling, was tiring following a prolonged pursuit of Denny Hulme's Brabham-Repco, which subsequently went on to win the race.

In the closing stages, Bandini's Ferrari 312 clipped the right-hand barrier as it negotiated the waterfront chicane, slewed out of control, and mounted the straw bales on the left-hand side of the circuit. The wayward Ferrari demolished a lamp standard, rolled and erupted in flames.

Bandini was eventually extricated from the inferno, but succumbed to multiple injuries three days later. The whole gruesome affair was the subject of world-wide television coverage, and, rightly, was responsible for accelerating the development of a safety-orientated approach to the business of international motor racing.

The Bandini tragedy was followed closely by more major disasters in 1968. The great Jim Clark was killed at Hockenheim when his Lotus 48 F2 car suffered a puncture which hurled it into a forest of unprotected trees at the trackside. Mike Spence died from head injuries testing at Indianapolis. Jo Schlesser was burnt to death when his Honda crashed at Rouen-les-Essarts in the French Grand Prix.

In 1969, Lucien Bianchi died when his Alfa Romeo T33 sports car crashed and burned during pre-race tests at Le Mans. At the end of 1970, World Champion elect Jochen Rindt died from multiple injuries when his Lotus 72 crashed during practice for the Italian Grand Prix at Monza. It was the sort of incident from which, had it involved a state of the art carbon fibre composite chassis two decades later, the driver would probably have walked away.

By the time the popular Swiss driver Jo Siffert was asphyxiated in his burning BRM P160 in October 1971, having survived an initial crash impact which left him only with a broken leg, the momentum for dramatically enhanced constructional safety was building fast.

The previous six years had seen a fundamental change in attitudes, prompted largely by the crusade waged by Jackie Stewart. He emphasized the view that racing drivers were paid for their skill, not purely to subject themselves to the risk of death and injury.

It is a measure of how attitudes have changed over the past 25 years that Stewart's views were regarded as little short of heresy in some quarters. Today, the precept that safety should be of paramount consideration is absolutely taken as read.

Although Jackie Stewart drove one of the superbly engineered Cosworth-Ford-engined Matras to victory in the 1969 World Championship, Team Lotus bounced back in 1970 with the superb Lotus 72, and Ferrari revitalized its challenge with the splendid flat 12-cylinder 312B. However, a close analysis of some of these machines reminds us that Grand Prix cars were still very frail pieces of equipment by the standards we have come to know in the 1980s, structurally questionable, and prone to catching fire as the result of relatively minor impacts.

Chapman's Lotus 72 pioneered the hip-mounted water radiator position, a configuration which enhanced weight distribution while at the same time removing the possibility of an overheating footwell giving rise to driver discomfort. Chapman's design also included an aerodynamic chisel nose profile and the use of torsion bars as a springing medium.

However, the car's Achilles' heel proved to be its use of inboard-mounted disc brakes, in an effort to reduce further its unsprung weight, and it was almost certainly the failure of a front brake driveshaft which caused Jochen Rindt's fatal accident. Even more worryingly, the manner in which the Lotus 72 monocoque was ripped asunder in what, even to pursuing drivers, looked at the time like an extremely modest impact, served as a reminder that much more work needed to be done in terms of structural safety if Formula 1 was not to continue killing its drivers with an unconscionable frequency.

The late 1960s and early '70s, of course, saw the heyday of the circuit safety crusade, spearheaded largely by Jackie Stewart. This does not strictly fall within the terms of this book, but as a barometer of contemporary feeling this movement presaged significantly improved F1 constructional safety standards by two or three years. By the early 1970s, the circuits had become significantly safer, but the machinery itself was still lagging behind.

Prior to Rindt's accident, the 1970 season had also been highlighted by two more massive fires, one of which had killed the popular British driver Piers Courage, when the Frank Williams team de Tomaso crashed during the Dutch Grand Prix at Zandvoort. Potentially even more serious, however, was the accident which befell Jacky Ickx's Ferrari 312B on the opening lap of the Spanish Grand Prix at Jarama.

Its aluminium fuel tanks brimful, and extending to the very outer perimeter of the monocoque, the Ferrari was T-boned by Jack Oliver's BRM P153, after the British car suffered a stub axle failure while underbraking for a tight downhill left-hand corner. The Ferrari's frail outer monocoque skin was punctured, and the two cars erupted into a fireball from which both drivers were extremely fortunate to escape. This powder-keg fragility was worrying in the extreme, and the ever-present fire danger increasingly moved towards the top of the safety agenda in the early 1970s.

At the end of the following year, Jo Siffert's fatal accident at Brands Hatch threw up renewed concern over safety facilities. Yet despite experiments with safety fuel tanks, it was not until the start of the European Grand Prix season in 1973 that deformable structures cladding the sides of the monocoque were made mandatory in an attempt to provide additional impact resistance to the area covering the fuel tanks.

Taking the safety campaign through its next logical step, was the introduction of rubberized safety fuel cells. Even this development could not provide total protection from major impacts, as the fiery accident to Niki Lauda's Ferrari 312T2 during the 1976 German Grand Prix at Nürburgring so graphically reminded us. But with the advent of ground-effect aerodynamics, the need for uncluttered sidepods in the interests of clean airflow along the sides of the cars effectively demanded the incorporation of a single, central fuel cell mounted immediately behind the driver. This safety development, even though it was born from the requirements of enhanced performance, was extremely significant.

The 1970s were largely dominated by Cosworth-Ford V8-engined machines, although Ferrari enjoyed an outstanding run from 1975 through to '79, winning three out of five Drivers' World Championships. On the technical front, Maranello pioneered the use of the transverse gearbox mounted ahead of the rear axle line on their 1975 title-winning 312T. This effectively processed a concept which had been tried fleetingly by March designer Robin Herd on the Bicester team's type 721X in 1972, but this system was based round a ponderous Alfa Romeo gearbox (inboard, but not transverse) which was borrowed from the T33 V8 sports car, and was blighted by a variety of secondary problems which made the viability of the concept extremely difficult to assess. Today, the incorporation of a transverse gearbox, mounted ahead of the differential, is an accepted element in the design of every Grand Prix car.

Finally, in 1980/81, McLaren and Lotus signalled the dawn of a new era with the adoption of carbon fibre composite construction, opening a whole new dimension to the business of Grand Prix car technology which is dealt with in specific detail within this volume.

CHAPTER 1

GENERAL ENGINEERING

In the summer of 1967, Jack Brabham's Formula 1 team introduced a new Grand Prix car. The Brabham BT24 was a refined version of the machine which the previous year had carried the tough, pragmatic Australian to the last of his three World Championships. As it transpired, the BT24 sustained the Brabham-Repco pedigree in fine style. No matter that its maximum rear track dimensions were dictated by the width of the team's transporter at the time.

This tale is told not to decry the Brabham team's achievements, which were considerable, but to make a subtle point. To wit, Grand Prix motor racing in the 1960s may have looked from the touchlines as though it was a technically sophisticated affair, but technical improvisation was frequently more critical in achieving success than sheer engineering genius.

The coming of the 1950s really closed the door on the concept of the 'big manufacturers' becoming involved in motor racing, in the sense that they built the cars in addition to owning and operating the teams which raced them. It matters little, for the purposes of this narrative, that Grand Prix racing has frequently relied on the small specialists, the little men with the genius for getting things done, to provide real strength in depth: just as Tony Vandervell would rely on input from experts such as Colin Chapman and Frank Costin to fettle his Vanwalls into World Championship challengers, just as Ford relied on Cosworth Engineering to put their name on the Grand Prix map. Almost all the way through the story of post-war Grand Prix racing it was the small-time operators who held the key, either as technical catalysts or in the role of second division teams and drivers who did so much to flesh out the starting grids.

In the late 1950s, the Cooper, *pere et fils*, began installing what had originally been designed as a fire pump engine in their Grand Prix cars. The big four-cylinder Coventry-Climax unit thumped its way to World Championship victories in 1959 and '60, after which Climax introduced a V8 for the 1.5 litre F1 regulations, allowing every man and his dog – or so it seemed – the chance to have a crack at the Grand Prix game.

It was against this racing backdrop that the expression 'kit car manufacturer' was originally coined. It was hardly a neutral label, more a disparaging reference to racing teams who, so the critics believed, owed such capabilities as they had to the blacksmith's art rather than that of the true automotive engineer.

Throughout this period, Ferrari was held up as an example to others. Now *there* was a racing team. It designed, built and raced its own cars in many international categories. As did the British BRM organization. Yet between 1962 and 1965, the combination of the mercurial Jim Clark and his Coventry-Climax-engined Lotus 'kit car' designed by Colin Chapman was, overwhelmingly, the class of the field. Small may not have been beautiful in the eyes of the die-hards, but it was certainly more successful.

Two decades later, the 'kit car' makers were still at it. At least, that's how it looked from a distance. But to compare the notoriously dingy, damp and dark Cooper headquarters at Surbiton, or the cramped environment at Lotus's Cheshunt factory, by then literally bursting at the seams, with the clinical, orderly and unfussed environment which now prevailed at McLaren, Williams or Benetton made one understand that they were no longer special builders in the pejorative sense of the word.

Now the engine manufacturers beat a path to their door. In the case of McLaren, the team was so well funded that it could afford to commission Porsche to manufacture a bespoke twin turbocharged V6 engine to absolutely, and precisely, the specification laid down by the then McLaren Technical Director John Barnard.

In the same way as the Wright brothers' first flight could have been accommodated within the fuselage of a Boeing 747, those pioneering aviators would have been awestruck had they lived long enough to see the four-engine behemoth which represents today's state of the art long-distance airliner. But such has been the pace of motor racing technology that the transition from Cooper-Climax to McLaren-Porsche took barely two decades, and there were plenty of old hands from the previous era to stand around and marvel at the high technology offerings of the '80s and '90s.

Think about the cars that lined up to contest the first year of the 3 litre Formula 1 regulations in 1966. Much fancied was the 3 litre Ferrari 312, but only on the basis that it had been the first such machine to be readied. Its announcement was surrounded by an aura of almost mystical anticipation. The bottom line, however, was that the car was an unwieldy dog: even its engine was borrowed from cars that had gone before.

Consider John Surtees's early season observations about the machine after an intensive test programme in the little Tasman Dino 246 which had lain unused over the previous winter in the wake of his near fatal accident in the Can Am Lola T70 sports car at Mosport Park the previous autumn:

'I'd been round and round Modena in that little Tasman 246 and then, of course, the V12 was wheeled out. Everybody was saying "Ah, 3 litre formula, Ferrari will walk it, foregone conclusion, Surtees has got no problems..." So I take this thing out and it's two and a half seconds slower round Modena than the V6. Flat as a bloody pancake! It transpires that all it is is a sports car engine reduced to 3 litres. And everybody's saying "Poor old Jack Brabham has only got 290/300 bhp from his Repco engine." This bloody V12, which weighed God knows how much, was really only giving 270 bhp... I went on to win the first race at Syracuse, but had to row it along like hell.'

Despite Surtees's protestations, the Ferrari team seemed optimistic and upbeat. But although the spaceframe Brabham with its production-based light alloy Repco engine had been present at Syracuse, minor technical problems had prevented it from demonstrating its true potential. But come the non-title Silverstone International Trophy and Brabham simply flew away from Surtees, leaving the Ferrari trailing by over nine seconds at the chequered flag.

In the wake of this result, there was some friction back in the Ferrari engine department – and not of the mechanical kind. Despite Chief Engineer Mauro Forghieri's assertions that Surtees was off-beam with regard to his quoted power output figures for the V12, John nevertheless told the Italian engineer, and engine man Franco Rocchi, precisely what he thought of the gutless V12.

'They went very quiet', recalls the Englishman. 'Then

they said "Well, of course, we can't afford to build a totally new engine, you know. We've got to streamline things a bit, use parts which are interchangeable with the sports car engines". Later in the season Surtees would quit the team as a result of a dispute with team manager Eugenio Dragoni over the question of driver pairings for Le Mans. Having lost their sole top-line driver, the Ferrari V12 was thereafter cast in the role of pathetic also-ran, apart from at Monza where Italian Ludovico Scarfiotti kept his head to score a rare home victory.

Meanwhile, Jack Brabham's homespun spaceframe Brabham-Repco must have been one of the least likely machines ever to have raced to a World Championship title. By the early 1960s, the Melbourne-based Repco company – Replacement Parts Pty Ltd – had developed into the largest manufacturer of automotive components to the Australian vehicle manufacturing industry. Repco made pistons, piston rings, bearings, ring gears, gaskets, clutches, drive-shaft assemblies, drop forgings and brake components, as well as having a thriving business supplying hand and machine tools, dies for metal pressings, gauges and balancing machines.

At the time, Australasia's most prestigious single-seater racing category was the Tasman Series, the lifeblood of which had been the long-stroke, four-cylinder Coventry-Climax FPF engine. This unit had a respected pedigree, but it was clear that supplies of spares for these engines were not open-ended. Thus Repco's Chief Engineer Frank Hallam, working with Project Engineer Phil Irving, developed a new V8 which was based round the General Motors Oldsmobile F85 cylinder block, an engine which had effectively been abandoned by GM after initially being evolved as part of a linerless aluminium engine programme for a projected 3.5 litre compact Buick sedan planned for the US market.

Developed from such an unlikely source initially as a Tasman formula engine, Jack Brabham got wind of the project and persuaded Repco to stretch it to 3 litres for the new Grand Prix engine regulations. The first engine was then installed in an interim Brabham BT19 spaceframe chassis which had originally been earmarked for the still-born 1.5 litre Climax flat 16 engine. The net result was a technical combination which won four Grands Prix and the World Championship for Brabham at the age of 40 in 1966.

Repco developed the engine effectively enough for Brabham's team-mate Denny Hulme to bag the Championship in 1967, but for 1968 a four-cam version of the Australian V8 proved dramatically unreliable, failure after failure adding up to a catastrophically disappointing year for Jack and his new running-mate Jochen Rindt.

Just keeping the Brabham-Repcos running throughout that troubled summer was an exercise in what now seems like stone-age technology. Two of the most graphic tales of improvisation are recalled by John Judd, then working as Brabham's engine man with Repco, and now one of the leading specialist race engine manufacturers in Britain:

'I'd spent much of 1967 working down in Australia on the four-cam engine. The power output was OK, but when it came to racing we had a large number of quality control problems. Gudgeon pins, for example... We spent one weekend rushing around converting the engines to take gudgeon pins out of a Petter diesel engine. We also had difficulties with the valve seats. We went down to Jarama for the Spanish Grand Prix, the second race of 1968, and Jack had a failure caused by a valve seat dropping out, so he had to scratch and Jochen started from the pit lane because we were all having teas in the transporter, not quite knowing when the race was due to start, and then we heard the cars going out on the warming up lap.

'Of course, Jochen got browned off with the whole situation, but he knew we were working bloody hard on that engine. Eventually we worked out that Repco was making the valve seats out of the wrong material which caused them to shrink. So the night prior to the Belgian Grand Prix we found ourselves tearing down one of those V8 engines after it had shown the first sign of trouble during practice at Spa.

'Jack flew it home after Saturday practice, and although we didn't have the right equipment at the factory, Ron Cousins, who used to work with HRG, single-handedly drilled out the old valve seats on a radial drill, made new ones, put them in, cut them, and the heads were then cooked in Betty Brabham's domestic oven. She woke up at about three o'clock in the morning and almost called the fire brigade because of all the fumes in the house. The engine was then reassembled and flown back out to Spa again sitting in the right-hand seat of Jack's Piper Twin Comanche!

'There was also the occasion when the cars had left for the Dutch Grand Prix at Zandvoort when we got a message through from Repco in Melbourne to say that the piston clearances in the most recently delivered engine was incorrect, and if we started them up the pistons would make contact with the valves. So Jack bought a chisel in a Guildford hardware store and we chiselled down the piston crowns when the cars got to Zandvoort. I didn't know enough, really. I was a bloody virgin...'

The 1966 switch to the 3 litre engine regulations had also seen the Cooper team switch to a Maserati V12, a real makeshift affair which was essentially a development of the 2.5 litre V12 which had seen service briefly in the front-engined Maserati 250F a decade earlier. It resulted in a heavy and cumbersome package which nevertheless proved remarkably effective.

Meanwhile, BRM would experiment with its disastrously over-complex H16 engine, a programme which endured for only two seasons before it was supplanted by a more conventional V12. The fledgeling McLaren team would become the first outside customer for the BRM V12 during the summer of 1967, having briefly experimented with unsuitable V8s from Ford (pushrod, stock block) and the Italian Serenissima company during the debut season the previous year.

It was perhaps typical that the innovative Colin Chapman was not prepared to be satisfied in the role of customer to BRM, a situation in which he found himself as the start of the 3 litre formula approached. During the 1.5 litre F1 era, the Lotus team had asserted itself as the number one Coventry-Climax performer, always benefiting from above average engines from this source. Now Chapman wanted to rekindle just that situation, so it was to Keith Duckworth and Mike Costin and Cosworth Engineering that he turned for salvation.

Chapman asked Duckworth whether he felt the Northampton-based company could produce a suitably competitive Formula 1 engine, and Keith answered in the affirmative. Chapman successfully canvassed Ford to provide the necessary financial support for the project, and Duckworth successfully produced the superb state of the art, four overhead camshaft 90-degree V8 – together with the four-cylinder FVA F2 engine – within an allotted budget of £100,000. Bearing in mind the contribution the Cosworth DFV would make to Grand Prix racing history over the next two decades, this must be regarded as arguably the Formula 1 bargain of all time.

The engine installation in the Lotus 49 was beautiful and uncluttered, with the V8 acting as a stressed member with the absolute minimum of accessories to clutter the purity of the basic design. At the rear, triangulated tubular frames provided inboard upper suspension pick-ups on the cylinder heads and a crossframe provided lower mounts for the reversed lower wishbones below the very light ZF five-speed gearbox. Suspension location also included a single top link, and twin radius arms feeding drive and braking loads into brackets positioned on the monocoque rear flanks. Graham Hill put the Lotus 49 on pole position for its race debut in the 1967 Dutch Grand Prix at Zandvoort, but it was his team-mate Jim Clark who whistled through to give the new Cosworth-engined machine a timely victory on its maiden outing. At a stroke, the parameters of contemporary Grand Prix car design had been redefined.

At this point, Grand Prix racing faced a potentially ticklish problem. As part of the deal for initiating the Ford-Cosworth development partnership which led to the DFV, Colin Chapman had negotiated a long-term contract ensuring Lotus's exclusive use of the engine.

Ford supremo Walter Hayes recalls that it was at the German Grand Prix of 1967 that he approached Chapman on the delicate matter of perhaps relinquishing that exclusivity. In a sense, the Cosworth-made V8 was proving too much of a success. Unless it was freely available to other teams – and other teams most certainly wanted it – then the whole business of F1 might collapse under the sustained ordeal of seemingly endless Lotus-Ford domination. Recalls Hayes:

'I said to Colin, "You know, if we don't make this engine available to everyone else, we'll kill the whole thing because there's no reason why anybody else will ever win." I've always said that I think it was a sign of extraordinary maturity by Chapman that he didn't even argue about it. He said something like, "Oh well, what a pity" and we agreed. It was as simple as that.'

In 1968, Matra and McLaren joined the Cosworth bandwagon and Jackie Stewart came very close to beating Graham Hill to the World Championship. Hill had now effectively assumed the burden of Lotus team leadership after the tragic death of Jim Clark in a minor league F2 accident at Hockenheim on 7 April that year.

Then came the technically confused 1969 season in which the structural vulnerability of spindly suspension-mounted aerofoils was put on public display when the Lotus 49Bs, driven by Hill and Jochen Rindt, both crashed heavily when their aerofoil mounts failed during the Spanish Grand Prix at Barcelona's spectacular Montjuich Park circuit.

Only by great good fortune were spectators not involved in their dozens and the sport's governing body, the CSI, was aghast. Acting in a unilateral manner that would be much more difficult today, or so we thought, under the provisions of the Concorde Agreement (the operational code by which F1 was governed since the early 1980s) they banned these high aerofoils at a stroke – the day between first and second qualifying sessions for the Monaco Grand Prix.

The 1969 season was also characterized by the development of a spate of four-wheel-drive F1 cars, a blind alley which was never subsequently pursued. Ironically, the man who probably did the most in terms of methodically and carefully examining the possible application of four-wheel-drive to 3 litre F1 cars was Cosworth DFV architect Keith Duckworth.

He made up his mind that four-wheel-drive was the answer after watching the 1967 German Grand Prix at Nurburgring. But he did not intend to design a system for another manufacturer to utilize. Instead Duckworth would plan his very own Grand Prix car, designing his own four-wheel drive system to provide what he anticipated would be the best and most effective method of transmitting the substantial power of the DFV through to the wheels.

Duckworth had already seen Chapman's Lotus 49 powered by his brainchild out-run the Repco, Weslake, Maserati, BRM and Ferrari opposition in the hands of the incomparable Jimmy Clark, yet he wasn't satisfied with the way in which the Lotus 49 put its power down on the track.

The early 49s suffered from terrific wheelspin out of corners, but the driving brilliance of the two talented Team Lotus men, Clark and Hill, allied to the enormous power advantage they had over their rivals, assured them sweeping success.

In 1967 ultra-wide racing tyres were in their infancy and the high aerofoils, which were to cause so much strife in 1968/69, had not made their shaky appearance on the F1 scene. It therefore looked as though four-wheel drive would be a timely and realistic technical proposition, and in the months after the 1968 Grand Prix season Duckworth finally completed his plans.

McLaren engineer Robin Herd left his successful career at the Colnbrook team's factory to join the project, fully appreciating that a spell at Cosworth would provide him with invaluable engineering experience. In the early months of 1969 the new Cosworth car carefully took shape at the firm's Northampton factory, revealing itself to be a striking and distinctively different Grand Prix machine.

The basis of Herd's design was a pair of sponsons between the wheels on each side joined by a stressed steel floor, the whole chassis achieving its rigidity from the engine and box-like structures front and rear. The side sponsons carried the fuel, the front and rear boxes containing the differentials as well as acting as pick-up points on which the suspension units were mounted. The finished package was clothed in a smooth, streamlined body with a wedge-profiled nose section which correctly anticipated trends in F1 over the next two seasons.

The four-wheel-driven Cosworth DFV motor – the only one to be cast totally in magnesium – was turned from its conventional position through 180-degree so that the clutch faced forward. From there a Cosworth gearbox containing Hewland gears with six forward speeds, the casing of which was actually built at Northampton, side-stepped power to an angled bevel central differential on the right which then shared the torque between the front and rear wheels.

The driver sat slightly offset to the left to allow for the fore/aft shafts running the length of the car, although the rearward shaft went only as far as the back of the engine, at which point a transfer box took it back to the centre and thence to a rear differential. By utilizing the drive-shafts, inboard disc brakes were employed front and rear.

'It was possibly all rather shallow thinking,' recalled Herd many years later. 'If we'd all thought a little bit longer, then we would probably have realized with the trends going the way they were that four-wheel drive wasn't practical.' By 'we' Herd meant Cosworth plus Lotus, Matra and McLaren, all of whom were anxious not to get left behind, and were frantically designing entries for the four-wheel-drive bandwagon.

However, Cosworth was arguably the first to realize its misjudgment. Although the company did not have its own racing team, they had intended to run the car in the 1969 British Grand Prix with former Lotus driver Trevor Taylor at the wheel. Instead, they withdrew the car after preliminary testing by Taylor and Mike Costin, himself a very capable driver.

There were three main problems, two of which were easily overcome, but the third remained a fundamental stumbling block. Firstly, in order to get the Cosworth's weight distribution right, Herd situated the oil tank just behind the driver, which proved too hot for comfort. The tank was subsequently moved to the rear, but then a front drive-shaft broke during testing. This was redesigned, but then came problems with the front differential.

With the terrific cornering forces generated by Formula 1 cars in 1969, difficulties immediately arose on fast cornering. In a right-hand corner, for example, the load was primarily on the left front tyre and there was hardly any load on the right. This meant it was prone to spin madly in tight corners, causing an aggravating degree of understeer as the front end tried to slide wide.

Of course, with the power as well as steering through the front wheels, the driver's job was made pretty strenuous. Duckworth and Herd reasoned that either very little torque should be put through the front wheels, or some sort of limited-slip differential should be incorporated into the design to minimize the wheelspin.

Taken to its extreme, the former course would have negated the whole purpose of four-wheel drive and, while the idea of a limited-slip differential was fine in concept, no suitable differential to perform the task was available.

Nevertheless, the Cosworth was duly tried with limited-slip differentials and even Jackie Stewart, testing his Formula 1 Matra at Silverstone one day, was persuaded to slip into Duckworth's machine for a few laps to offer an opinion. He came back after a short time complaining, 'Its so heavy on the front, you turn into a corner and the whole thing starts driving you. The car tries to take over.'

Stewart just about summed it up. But the other constructors were determined to race their costly new four-wheel-drive creations. For the 1969 Dutch Grand Prix, both Team Lotus and Ken Tyrrell's Equipe Matra International arrived with new cars in addition to their two-wheel-drive machines, Chapman's in particular being striking in the extreme, the Lotus 63 being a direct descendent of the Pratt & Whitney-powered turbine Indy cars and utilizing quite a number of components from those machines which are touched on elsewhere in this volume.

Last to be deflected from pursuing this technical avenue was Chapman at Team Lotus, despite implacable opposition to the project from both Rindt and Graham Hill. Eventually the Lotus 63 project was entrusted to the hands of the team's junior driver John Miles, and it ran respectably in several races. Rindt finally relented, driving a type 63 to second place in the non Championship Oulton Park Gold Cup, while Miles drove the car again in the Italian Grand Prix where Stewart again tried the Matra MS84 in practice.

Those remaining juggled with the torque split ratios of

their cars, gradually backing off so that only 30 per cent or so went through the front wheels. Eventually the whole four-wheel-drive Formula 1 honeymoon petered out inconclusively at the end of 1969, never to be repeated.

A few years later Colin Chapman summed up the way F1 subsequently evolved:

> 'We have achieved everything we can achieve with two-wheel drive and four-wheel drive now has no advantage to be offered. We achieved the subsequent lap speeds with no increase in weight which, however minimal, we would have had to carry in a four-wheel-drive car. And we didn't have to suffer any of their handling quirks either.'

Tony Rolt, then Managing Director of FF Developments, the Coventry-based company which specialized and pioneered many sophisticated four-wheel drive applications for road cars, concurred with Chapman's viewpoint.

He said in 1973, 'While Grand Prix cars continue to have such unrestricted tyres and can produce such terrific cornering forces, I cannot see four-wheel drive having a future in Grand Prix racing.' And that is where this particular story ended.

CHAPTER 2

CHASSIS TECHNOLOGY AND DESIGN

The demands of ground-effect F1 aerodynamics in the late 1970s called for the use of much stiffer, stronger materials from which to manufacture the chassis and with a much smaller cross-sectional area than hitherto. A glance at the central fuselage section of the Championship-winning 1978 Lotus 79 reveals this essential point. At the height of the ground-effect era, drivers would be strapped into this central 'tube', ahead of the fuel cell and engine, while as many of the car's technical ancillaries as possible would be packaged in such a way as to provide

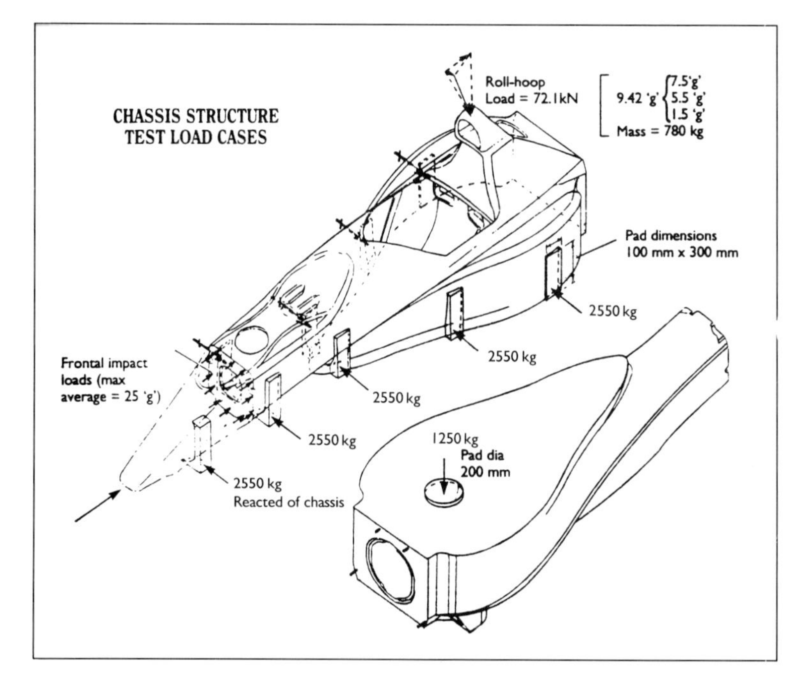

From the start of the 1992 Formula 1 season, each individual F1 chassis had to be subjected to a side-pad pressure test of 2550kg and a 1250kg pressure test on the underside of the fuel cell area, as shown by this diagram.

the minimum possible interference with the aerodynamic side pods, which were the key to the car's operational efficiency.

The structural loadings imposed on a Grand Prix car chassis were soon enhanced by a pressing need to minimize the weight penalty imposed by heavy and prodigiously powerful turbocharged engines, which helped to accelerate the thinking of many F1 designers with regard to using carbon fibre composite materials in monocoque construction during the late 1970s and early '80s.

By 1979, Brabham designer Gordon Murray had produced the Alfa Romeo V12-engined BT48 for the team, at that time owned by Bernie Ecclestone. This was the first contemporary Grand Prix car to make significant use of carbon fibre composite panels in its monocoque construction.

Before that, the use of composite materials had been conducted on a somewhat patchy and experimental basis. During the Second World War, the development of fibreglass and polyester matrix resins was prompted by a requirement to produce such components as lightweight protective domes for radar equipment fitted to aircraft and warships. In addition, wartime research work also led to the development of adhesives capable of metal to metal bonding to aerospace standards. The earliest recorded composite aircraft structure was that of a Spitfire fuselage. Produced in anticipation of a severe aluminium shortage early in the war, it consisted of skins made from flax fibres impregnated with a modified phenolic resin, which was then cured under heat and pressure.

This development was produced by Aero Research of Duxford, near Cambridge, which is now part of Ciba-Geigy Bonded Structures Ltd. Although the aluminium shortage did not materialize, and the fuselage was never incorporated into a completed aircraft, tests at Farnborough's Royal Aeronautical Establishment revealed it to meet all its required specifications.

The motor racing fraternity had been familiar with glass fibre-reinforced plastic ever since the 1950s, when it had been used for the manufacture of body panels for both racing and road-going sports cars. 'Wet lay up' fabrication, as the technique came to be known, involved applying a coat of resin by brush or spray to a mould which had been pre-treated with a release agent, followed by a layer of glass fibre.

In those days, what was known as the 'chopped strand' method of manufacture was employed, that is to say, the strands were 'laid up' in random directions, the theory being that this would make the resultant material equally strong in all directions. This had secondary implications, since in order to gain the necessary strength, the high number of strands required could result in excessive weight burden.

This material facilitated the relatively cheap production of the complex compound curvature bodywork which increasingly replaced the hand-beaten aluminium panels clothing spaceframe racing car chassis at the turn of the 1960s. Still later, of course, Colin Chapman's pioneering design work led to the monocoque construction Lotus 25 which set new standards of Grand Prix car design in 1962.

However, while Chapman's creation was a straightforward aluminium 'bathtub' monocoque, essentially two aluminium box sections containing fuel tanks linked by a floor pan, at around the same time the Cooper Car Company's chief designer Owen Maddock came up with what must be considered the first contemporary composite F1 chassis, in preparation for the 1963 season.

This was an oval-shaped structure with an upper aperture through which the driver gained access, and twin booms at the rear to locate the engine. It was built with an aluminium outer skin, on the inside of which was bonded glass fibre. It was intended to accommodate the stillborn flat 16 Coventry-Climax engine which had originally been envisaged for the final year of the 1.5 litre F1 in 1965.

Sadly, it was never built up into a completed racing car. It was extremely still, not unduly heavy, but very costly to manufacture. The Cooper team was in decline and Owen Maddock was on the point of leaving. Cooper old-timers recall this composite monocoque being so strong that when it was finally consigned to scrap it proved almost impossible to destroy.

Carbon fibre origins

During the post-war years, Government-sponsored research programmes in the USA, Japan and the United Kingdom subsequently focused on examining the development of new, ultra-stiff and lightweight materials for use in an aerospace application. This work had originally spawned the use of carbon fibre composites, and the material first attracted wider public attention when Rolls-Royce used it for the first stage compressor fans in their RB211 jet engine.

Indirectly, the RB211 saga was to result in Rolls-Royce ending up in the hands of the receivers. These compressor fans worked well enough and proved structurally sound, but when a small amount of sand or other abrasive material was introduced into the airflow through the engines, it stripped the epoxy resin from the fibres and weakened the blades, leading in due course to failures.

Carbon fibre was used in the wing structures of several F1 cars during the mid-1970s, but the failure under aerodynamic loading of a carbon fibre composite wing support strut manufactured from this material caused a tragic accident in the 1975 Spanish Grand Prix. Whilst leading this race at Barcelona's Montjuich Park circuit, Rolf Stommelen's Hill-Cosworth vaulted a barrier and killed four onlookers. Several less serious incidents followed, causing designers to become understandably concerned about the use of these materials, and correspondingly cautious when it came to furthering their application.

In the same way as there had been doubts about the structural validity of the aluminium honeycomb employed in monocoque manufacture during the mid-1970s, so there were reservations about the use of carbon fibre. In the case of the former, concern had centred on the possibility

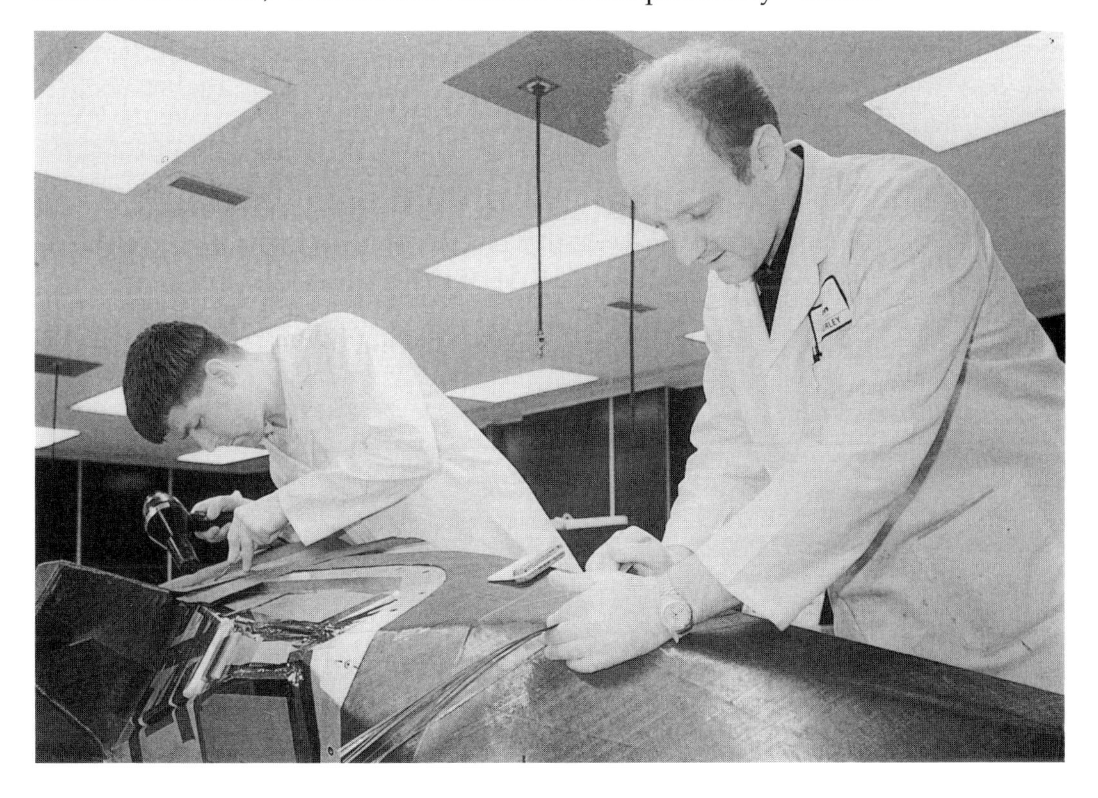

The Woking-based McLaren International team pioneered the use of carbon fibre composite chassis manufacture. Here two technicians lay up the carbon fibre for one of the team's Honda-engined MP4/6 chassis.

that the material might tear abruptly under sudden impacts, and that the glue used for bonding the outer skins to the inner honeycomb core would break down, allowing the chassis itself to flex.

In the case of the latter, concern centred on whether the carbon fibre sheet would simply shatter when subjected to the destructive forces involved in a major accident. As it transpired, carbon fibre composite chassis technology developed over the next decade to the point where it enabled more efficient, safer and impact resisting racing cars to be developed than could have ever been constructed using aluminium alloy materials.

The paradox, of course, is that carbon fibre has no inherent strength in its raw unprocessed form. It is originally produced in bundles of up to 12,000 fine filaments, each one a few thousandths of an inch thick, derived from specially treated acrylic fibre which has been oxidized by heating under tension in an oxygen-rich environment.

The end result is a material which is not isotropic, that is to say not one whose properties are directionally uniform. In much the same way as wood, the strength and stiffness are very high along the axis of the fibres, but relatively low across them.

Such fibres are relatively easily damaged in their raw form, and it is essential to avoid surface abrasions as these give rise to cracks in the finished structure which can induce failure.

These fibres have no high performance qualities until saturated with a small amount of resin and oven-cured at around 248-degree F (120-degree C), after which they become transformed into the stiff and light material which forms the basis of current advanced composite technology.

Unlike the chopped strand method of laying up glass fibre-reinforced plastic, the nature of carbon fibre required the designer to start with a significantly more precise idea of how he wished the final product to perform. Williams Team Technical Director Patrick Head explains:

'The secret, of course, is to know in which direction the load is to be applied, or the areas of greatest stress and strain within the monocoque, so you can lay the fibres correctly along the anticipated load paths.

'It should also be remembered that during the early days of carbon fibre composite in a motor racing application, there was no enormous fund of accumulated technical experience on which to draw, so far as Williams was concerned at least.

One of the reasons for the slow introduction of such technology was that we simply had insufficient knowledge about the material.'

Barnard's pioneering work

Late in 1979, Patrick Head put Ron Dennis in touch with John Barnard, ostensibly to discuss the design of an F2 chassis for Dennis's Project Four team. In fact, Ron was already thinking in more ambitious terms, laying the groundwork for a move into F1, at the same time as Barnard had been mulling over plans to build a Grand Prix chassis from carbon fibre.

By happy coincidence the two men got together, and aided by an extremely fortuitous set of circumstances, they found themselves steering the fortunes of the newly established McLaren International team less than 18 months later.

Barnard was then in a position to develop his pet theories. However, he was to be disappointed by the initial reception in the UK which greeted his ideas for a carbon fibre monocoque. 'To be honest', he recalls, 'there was not really a great deal of enthusiasm or interest at all. The big concerns who were capable of carrying out the work simply did not have the facilities to make such specialist components.'

By good fortune, a racing colleague of Barnard's from the USA was in a position to assist. This was Steve Nich-

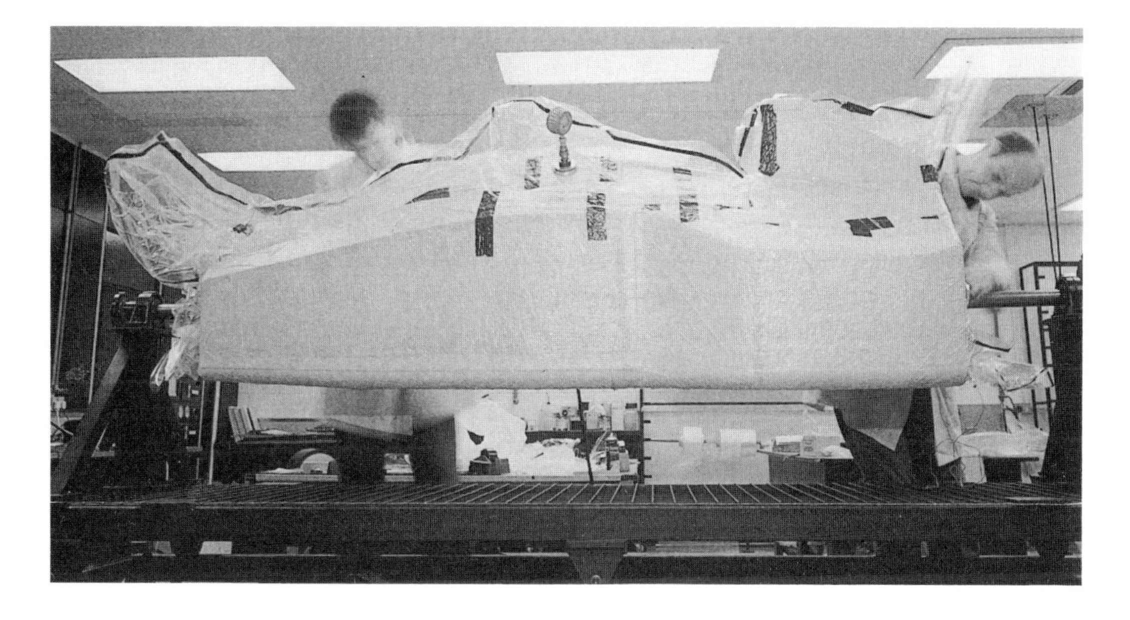

A McLaren MP4/6 chassis inserted in its vacuum bag for the purposes of curing the carbon fibre structure.

ols, from 1989 to '92 the chief chassis designer for the Ferrari F1 team. He suggested that Hercules Aerospace in Salt Lake City might be interested.

One evening Barnard returned to the McLaren factory at the end of another fruitless trip to a potential British supplier, and sat in the office as co-director Ron Dennis called the US company and outlined his teams requirements. Within a few days, Dennis and Barnard flew to Salt Lake City to explain McLaren's detailed needs and duly concluded the deal.

In terms of mechanical properties, Barnard rightly judged that the aluminium-skinned honeycomb monocoque concept had been exploited virtually to the limit of its performance. He was now seeking a major improvement in torsional chassis stiffness combined with a substantial reduction in weight.

'When we built the first carbon fibre MP4 chassis, we ended up with a structure with 15,000 ft/lb stiffness at a time when a conventional chassis had about 8,000 ft/lb stiffness, but the problem was that it weighed pretty much the same as an aluminium monocoque. Thereafter I reduced the CFC lay-up, which reduced the torsional stiffness to around 12,000 ft/lb, which was still a considerable improvement, but the weight came down by about 25 lb from a total of about 90 lb, so we made over 25 per cent savings in this respect.'

Hercules was already very active in carbon fibre technology, being involved in a wide range of complicated and ambitious projects such as the manufacture of components for missile rocket motors. They duly agreed to make the requisite panels for the new Barnard-designed McLaren, designated MP4. These were then flown to the team's factory in Woking, Surrey, where they were bolted and glued together to form a light, rigid completed structure.

Another key benefit accrued from this form of chassis construction, namely a dramatic reduction in the number of components required to make the monocoque. In Barnard's estimation, as many as 50 aluminium sections would be required to manufacture a conventional chassis, but the carbon fibre design of the first MP4 called for a mere five main mouldings, plus the separate outer shell.

Hercules was also able to support the McLaren venture by providing finite element analysis expertise to help Barnard's design process, and the US company actually undertook the task of assembling the initial monocoques.

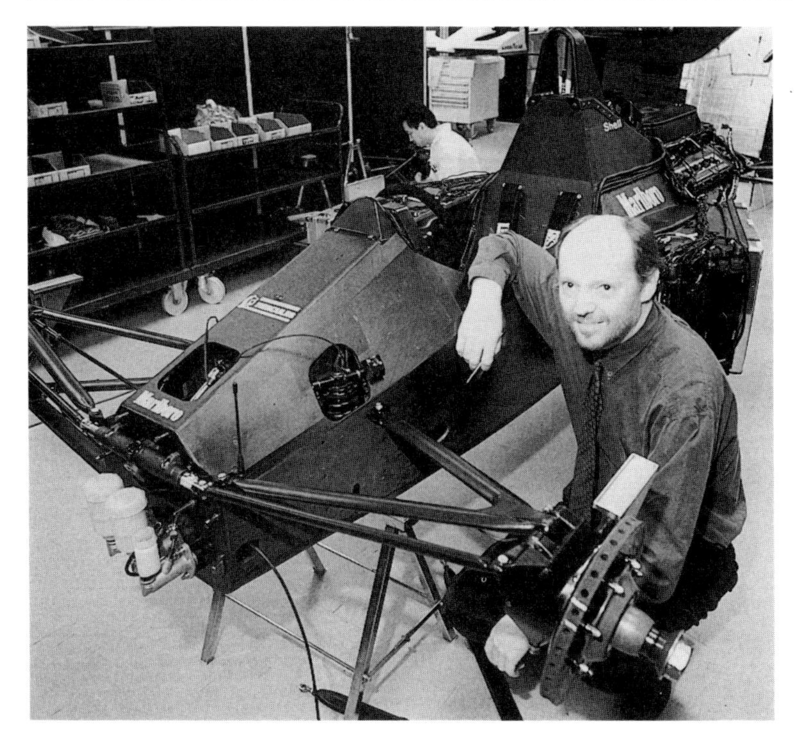

McLaren International engineer Steve Hallam poses with a completed MP4/6 monocoque, showing the relatively flat surfaces of the carbon fibre panels which go to make up the chassis. Removable body panels were fitted atop this structure, but further advances in computer-aided design technology enabled this car's successor, the MP4/7A, to have the monocoque shaped to more precise aerodynamic requirements, considerably reducing the need for removable upper panels.

The chassis of the McLaren MP4 was fabricated from high modulus unidirectional pre-preg and aluminium honeycomb, the components being bonded together on machined, cast aluminium tooling. The moulding was then vacuum bagged and cured in an autoclave oven.

The main monocoque structure included both the front and rear bulkhead, the only aperture being the cockpit through which the mould tool was extracted in sections. The other key components included the seat back, which enclosed the forward edge of the fuel cell area, an aluminium bulkhead which carried the front suspension components, inner panels forming the cockpit walls which also provided a box section strengthening of the cockpit area, and a dash panel bulkhead. The all-up weight of the completed chassis was around half the weight of the last aluminium McLaren monocoque.

Within a week of the McLaren MP4's launch in the spring of 1981, Lotus boss Colin Chapman harnessed all his powers of gamesmanship to announce that the monocoque construction of the new twin chassis Lotus 88, which effectively eclipsed Barnard's new machine in terms of carbon fibre technology by using a blend of carbon fibre and Kevlar, a man-made material manufactured by the US-based Du Pont Chemicals corporation.

Kevlar was based on an organic polymer, similar to a high-grade cellulose-like flax, but about four times stronger. Pre-impregnated with the crucial epoxy resin, carbon fibre/Kevlar cloth was commercially available and Lotus employed it for the manufacture of the type 88 monocoque.

The starting point was a flat sheet, about 8 ft (2.4 m) square, of carbon fibre/Kevlar-filled sandwich material within which was a honeycomb filling of Nomex paper foil. The epoxy resin was then applied by hand, after which Lotus left the sheet to cure chemically. Before the resin had totally hardened, pre-positioned apertures were cut into the structure in order to take mounting bobbins for engine pick-ups, sidepod fixings and suspension mounting components.

The whole panel was then scored and literally folded up around the dummy internal bulkheads, the edges bonded and taped together and the resin then left to complete its hardening process. The tube was then stiffened by the insertion of aluminium bulkheads, bolted in place to provide additional support for the engine and suspension mountings.

The end result was an outstandingly strong and durable structure, variations of which served Lotus extremely well for five seasons, and were superseded only by the advent of the 1986 Lotus 98T which was built round a one-piece moulded carbon fibre monocoque, in line with the trend of contemporary development.

This is one of the Williams team's autoclaves used for curing carbon fibre composite F1 monocoque structures, typical of the chassis manufacturing requirements demanded by a top F1 team in today's competitive environment.

Ferrari, Brabham and Williams would subsequently climb aboard the carbon fibre bandwagon during the mid-1980s, since when this form of chassis construction had become absolutely *de rigeur* for F1 applications and beyond.

However, at the height of the F1 turbocharged engine era in 1986-87, with power outputs leaping to a maximum of 1,200 bhp in high-boost qualifying trim, potential top speeds in the order of 200 mph (322 km/h) and cornering force load factors in the region of 3 G were being developed. It would thus become a prime requirement that an F1 monocoque should be imbued with tremendous stiffness in order that all the suspension and chassis components should interact with optimum efficiency.

The next significant step forward in carbon fibre chassis construction had come in 1983 when Gustav Brunner, then chief designer for the ATS-BMW team, designed a monocoque fabricated inside female composite tooling, the two halves of the structure being joined at the centre line. Following in the same vein, Patrick Head would adopt a similar manufacturing technique for the first moulded carbon fibre Williams FW10 in 1985. This first carbon fibre composite Williams chassis demonstrated a 60 per cent increase in torsional stiffness in the cockpit bay area when measured for the same weight of component.

As regards the relative benefits of male moulding and female moulding for monocoque manufacture, both systems have their advantages and disadvantages. These are

The Woking, Surrey, headquarters of the McLaren International team was opened for business at the end of 1987 and has come to be regarded as one of the most impressive of all Grand Prix team bases.

set in perspective by Dr Gary Savage, Head of Materials Development for the McLaren International team, writing in 1991:

'Looking at the various chassis designs, it becomes evident that the male moulded monocoques are beginning to appear a little dated. The male moulded variety is formed over a metal mandrel which is dismantled and removed from within the monocoque once it has been autoclave-cured. Such chassis have a fat, angular appearance similar to the old aluminium honeycomb constructions. The sharp angles of the chassis make it difficult to retain continuity of fibre orientation, but have the advantage of being a single piece structure.

'A female moulded chassis is built in two halves and invariably requires a join in what is a highly stressed component. This type of chassis follows the sleek aerodynamic profiles of the car and lends itself more readily to the maintenance of fibre orientation due to its gentle curvature rather than the sharp angles of the male moulded type.'

He went on to emphasise that a female-moulded chassis is inherently stiffer than its male-moulded counterpart because of its smooth curvature and wider cross-sectional area. This permits a saving of weight by the reduction of the number of bulkheads and the need to carry only a light engine cover rather than the full body top. Moreover,

The McLaren International machine shop is equipped with expensive, often purpose-designed, computer controlled machinery for a wide range of component manufacture.

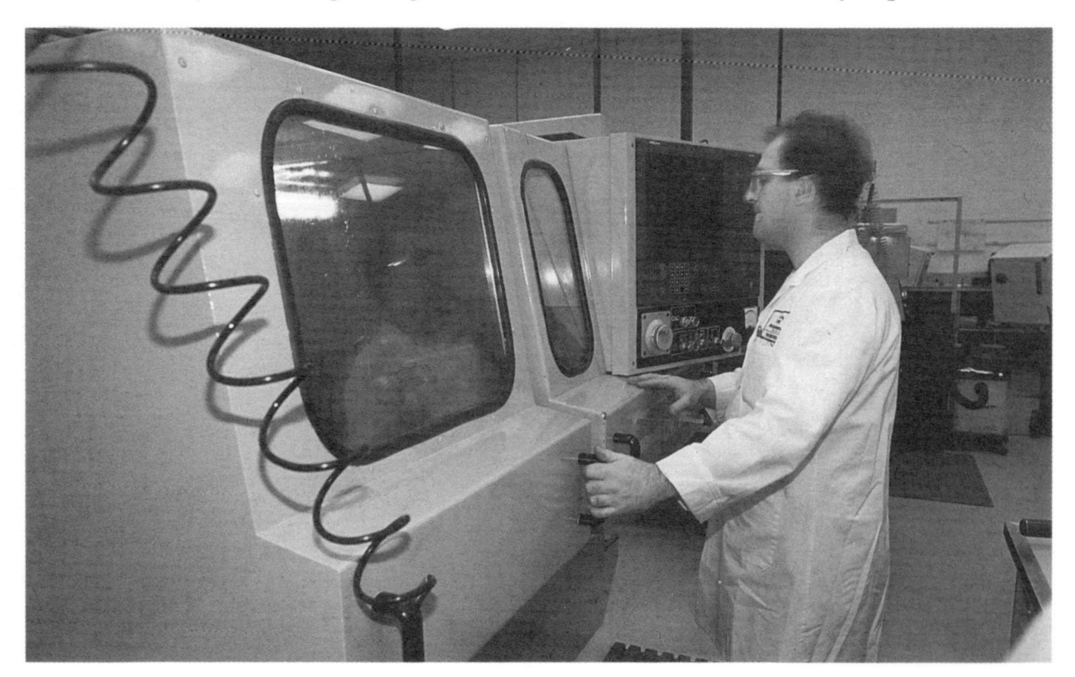

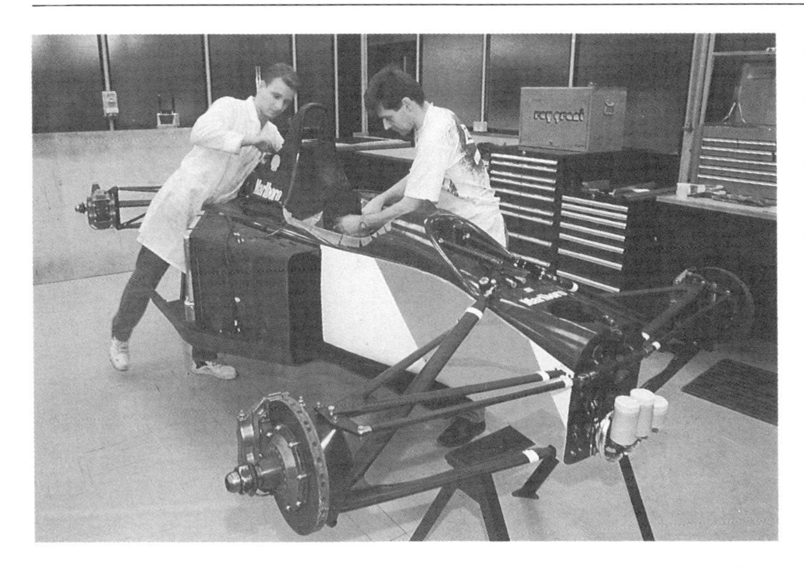

Race preparation bays at McLaren's base are spacious, light and airy – a far cry from the primitive conditions which were the norm in F1 less than two decades ago.

advances in adhesive technology now mean that rather than being a liability, the join may well be the strongest part of the monocoque.

Dr Savage also made the point that male-moulded monocoques were increasingly likely to be superseded by female-moulded structures, due to the ever-present need to reduce weight and the trend towards computer-controlled active suspension systems which obviate many of the procedures employed in setting up

McLaren International prides itself on the efficiency of its quality control department and has a scrupulous system of cross-referencing which can immediately identify the source of any component batch in the event of a problem arising.

conventional suspension systems.

Much of the monitoring and adjustment of these active systems is carried out by electronic remote control, requiring the mechanics to be involved in far less fiddling deep in the recesses of the chassis footwell, apart from initial installation, removal and maintenance. The need for access hatches and panels has thus been considerably reduced.

Ferrari make up ground

By the end of the 1970s, Ferrari was still basically constructing its monocoque chassis in much the same way as it had been doing a decade earlier; a basic lattice-work tubular frame overlaid with aluminium panelling for additional rigidity. By the turn of the decade the Prancing Horse was lagging behind badly, but quickly caught up by effectively plugging into British technology with the recruitment of former Hesketh, Wolf and Fittipaldi designer Harvey Postlethwaite in the summer of 1971.

By Postlethwaite's own admission, on his arrival, he would have liked to build a carbon fibre chassis immediately for 1982. However, the team seriously lacked the capability to do so, and he eventually advised Enzo Ferrari that if the team was going to tackle that sort of construction, it ought to be in a position to do it in house.

'We didn't want to find ourselves in a position where we subcontracted the work out', he recalls, 'so whilst we got ourselves organized on this front, I opted to build a honeycomb chassis on the basis that I knew how to put one together and it wouldn't fall apart.'

Since Maranello had no resin bonding facilities, Harvey utilized bonded Nomex honeycomb sheet manufactured by the Belgian Hexcel company for the Ferrari 126C2 chassis, a machine which closely resembled the last Wolf WR7 he had designed. The honeycomb sheet was folded round carbon fibre composite internal bulkheads and glued together, making a very strong and light structure.

The Ferrari 126C2 was used throughout the tragic 1982 season, in which Maranello won the Constructors' Championship despite the fact that Gilles Villeneuve was killed in practice for the Belgian Grand Prix at Zolder, and teammate Didier Pironi suffered multiple leg injuries during practice for the German race at Hockenheim, which finished his career.

The aluminium honeycomb C2 chassis was subsequently used through the first part of the 1983 season, with Postlethwaite's carbon fibre 126C3 appearing in time for the British Grand Prix at Silverstone, although not making its race debut until Hockenheim where Rene Arnoux won the German Grand Prix.

Ferrari was now firmly wedded to the carbon fibre composite route, and remained with Harvey Postlethwaite in charge through to the 1988 season, when John Barnard took over the technical reins at Maranello and his fellow Englishman moved to the Tyrrell team. What followed was the development of one of the most significant cars of the 3.5 litre F1 era.

Barnard and the Ferrari 640

By the middle of 1986, with the TAG/Porsche turbo-engined McLarens he had masterminded now on their way to a third consecutive World Championship, John Barnard concluded he was finding it difficult to sustain his motivation working with the team. He thus took a career decision which was to lead to the development of the most exciting F1 Ferrari in a decade, the type 640, a derivative of which would carry Frenchman Alain Prost to within sight of the 1990 World Championship title.

It was difficult to judge what had gone wrong in the personal chemistry at McLaren, but the working relationship between Barnard and Ron Dennis was on the verge of permanent derailment. In retrospect, it was a matter of two highly motivated individuals not having sufficient room to operate effectively beneath the same roof. They clashed, and the sparks flew. John later reflected:

'We had some simply terrible arguments, and sometimes the situation between us became positively explosive. I could feel myself tensing up, mentally revving right off the scale. Although things would periodically calm down and be OK, we definitely went through some frightful times together.'

Eventually Barnard left, but his design influence was to be seen on McLaren cars right through to the start of the following decade, retaining broadly the same monocoque shape as the original 1981 MP4 design. To this day, F1 McLarens are still made from carbon fibre materials manufactured by Hercules Aerospace. The cars continued to retain separate, removable upper bodywork through to the start of 1992, when the MP4/7 design signalled a significant reappraisal of McLaren F1 chassis construction techniques, and the links to the pioneering

The 1990 Ferrari 641 was the most successful development of the John Barnard-designed machine which had originally been conceived as early as the summer of 1988. Its tall, slim radiator ducts and sharply waisted side pods initially ducted all the cooling air out across the rear diffuser, but additional exit panels would eventually be opened in the outer wall of the pods as the V12 engine gradually developed more power and its cooling requirements increased.

MP4 were finally severed.

Unlike many of those over the years who have joined Ferrari's engineering staff, John Barnard was never besotted with the mystique or romanticism which to some extent has always surrounded the most famous Grand Prix team in the business. As a result, many of his Italian critics came to regard him as a reactionary, a newcomer from a foreign land who sought to impose his views and opinions on a not altogether sympathetic workforce when he took over as Technical Director in the autumn of 1986.

In reality, Barnard set out neither to antagonize nor to cosset. He was ruled by pragmatism: that, and the knowledge that he would be judged purely on what he achieved for the company. He was also recruited at a time when he was all too familiar with the level of competition ranged against Ferrari.

He was approached personally by Enzo Ferrari himself, the ageing patriarch following the team's periodic tradition of looking to England for technicians – or ideas – with which to pep up his Grand Prix team. But there were two elements of the deal with Barnard which looked as though they could be political dynamite.

Firstly, Barnard was to be invested with an unparalleled

degree of control on the engineering side, answerable only to the *Commendatore*. Secondly, he flatly refused to relocate to Italy and was therefore granted permission to establish his own engineering outstation, Guildford Technical Office (GTO), close to his Surrey home.

Although Barnard would leave the company in 1989, the most fascinating element of the whole affair was that Enzo Ferrari's strategy would be duplicated by the company's new President, Luca di Montezemolo, in 1992. By the end of that year, after a spell with Benetton and Toyota, Barnard was busy establishing a brand new design outstation in Britain as he prepared to craft a new generation of Grand Prix Ferraris.

Reflecting on his first stint with Ferrari, Barnard concedes that it did not prove as easy as he had perhaps imagined, in effect shifting the teams centre of gravity from Maranello to Guildford:

'You can look back on it and conclude that it might have been a mistake, and make a case that I should have worked in Maranello all the time. But if that had been what was on offer, I would not have joined Ferrari in the first place.

'I joined Ferrari in 1986 in good faith. I wanted to join because it was Ferrari, but perhaps I felt it could operate in rather the same way as we did at McLaren, which was a very structured company. You could split it up, take parts from here and there, independently. I assumed it would be possible to do the same at Ferrari.

'I had this fundamental theory that it was possible to fix almost any problem with the team provided you made a car which was quick enough and could win races. If you could go into a room, shut the door, and come out in six months time with a car that had no problems, and was quick, then I believed that 90 per cent of the problems would go away.'

Barnard found himself facing two challenges on his arrival. There was the short-term consideration of developing a turbocharged contender for the 1988 season, bearing in mind that this project was already well under way and, in the longer term, the more fundamental task of tackling an all new design for the 3.5 litre, naturally aspirated F1 which was due to come into full force at the start of 1989.

Cars for this new category would also be admissible to run alongside the 1.5 litre turbos in 1988 when the forced induction regulations would restrict these machines to a maximum fuel capacity of 150 litres and maximum boost pressure of 2.5 bar. Barnard knew that it would be a

The Benetton team's new technical headquarters at Enstone, near Oxford, had originally been designed and planned for the Bicester-based Reynard Racing Cars organisation and on its opening in 1992 eclipsed any facility owned by any rival Grand Prix team.

simply monumental task getting the necessary work done, let alone convincing the workforce that his way was ultimately the best for Ferrari's long-term interest.

Barnard's first task was to take over the development of the final 1.5 litre turbocharged machine, the F187, the design of which had been started by Austrian designer Gustav Brunner. The very ongoing nature of Grand Prix car design meant that he inherited a basic concept which he wouldn't have started with had he been in charge at the start. For Barnard, this was something of an anathema. This would not be the way as far as the new naturally aspirated car was concerned.

In fact, Ferrari would end the 1987 season on quite a buoyant note with Gerhard Berger taking the F187 to victories in the Japanese and Australian Grands Prix. For 1988, although it had originally been thought that Barnard's new creation might be ready, this proved an over-optimistic assumption and the team had no choice but to run a revised version of the turbo car – dubbed F1/87/88C – for the final year of the forced induction regulations.

This was a strategy which failed to reap any significant rewards. This was not because the Ferrari turbo was unable to outpace the first generation of 3.5 litre naturally aspirated cars. It managed that quite convincingly on most occasions. However, the McLaren-Honda partnership simply overwhelmed the entire opposition, winning a record 15 out of 16 races.

Eighteen months earlier, FISA's controversial President Jean-Marie Balestre had predicted confidently to a group of F1 team designers, 'I promise you gentleman, in 1988, there will be no chance for the turbos.'

Like many others, Balestre believed that a combination of a 150 litre fuel capacity maximum and a reduced 2.5 bar turbo boost pressure limit would render them uncompetitive. But the sport's masters had reckoned without Honda's considerable technical ingenuity, and their 1.5 litre turbos displayed an extraordinary blend of power and fuel efficiency such as stopped the opposition in its tracks.

Back at Maranello, however, Barnard had found 1988 to be fraught with problems. The first naturally aspirated Ferrari type 639 prototype had been readied for its first run at Fiorano in the hands of works driver Dario Benuzzi, in July, less than a month before Enzo Ferrari's death on 14 August .

Earlier in the season, the *Commendatore*'s son Piero Lardi-Ferrari relinquished his position as director of the racing team and moved to the road car division as Vice-Chairman of the company. Reportedly, Piero had found himself increasingly unable to adapt to what he regarded as Barnard's autocratic approach to the business of F1 engineering, so in the aftermath of his father's death the road seemed clear for Fiat appointees to begin dabbling in the Grand Prix business. As if F1 design procedures were not already complicated and challenging enough... John Barnard explains:

'Of course, after Mr Ferrari's death, there were very significant changes within the structure and hierarchy of the

The light, airy and spacious reception area at Benetton Formula's Enstone base.

company. Vittorio Ghidella was there at the time and seemed to feel that he had come to get involved in the racing side in a way that suggested he was going to replace Mr Ferrari.

'That caused a few problems, of course, although I suppose by that time I had started to sort out the people who were leaving the organization and was better able to get to the point where I could have contact and feedback with GTO.'

The arrival of Pier-Giorgio Cappelli at the Maranello helm briefly improved the volatile situation, but Barnard was again up against internal politics when it came to finalizing the design of the definitive naturally aspirated car. In this respect, the biggest sticking point was to become the principal feature which would ensure the car broke fresh ground, namely the adoption of the electro-hydraulic gear change system for the seven-speed longitudinal transmission which was under development for the new machine.

Happily, despite an enormous amount of opposition from some factions within Maranello, Barnard got his own way and the type 639 prototype remained firmly wedded to the semi-automatic change system.

As one of the first F1 teams to employ computerized finite element analysis techniques, Barnard and his design team expended an enormous amount of time and attention to detail on the design of the new Ferrari. This was the first all new F1 chassis Barnard had tackled since the original McLaren MP4 of 1981, that design enduring in service long after his tenure with the Woking team came to an end.

Working with carbon composite materials calls for a strictly controlled environment, both from the viewpoint of temperature and air quality. This is the 'clean room' specifically dedicated to this work at the Benetton headquarters.

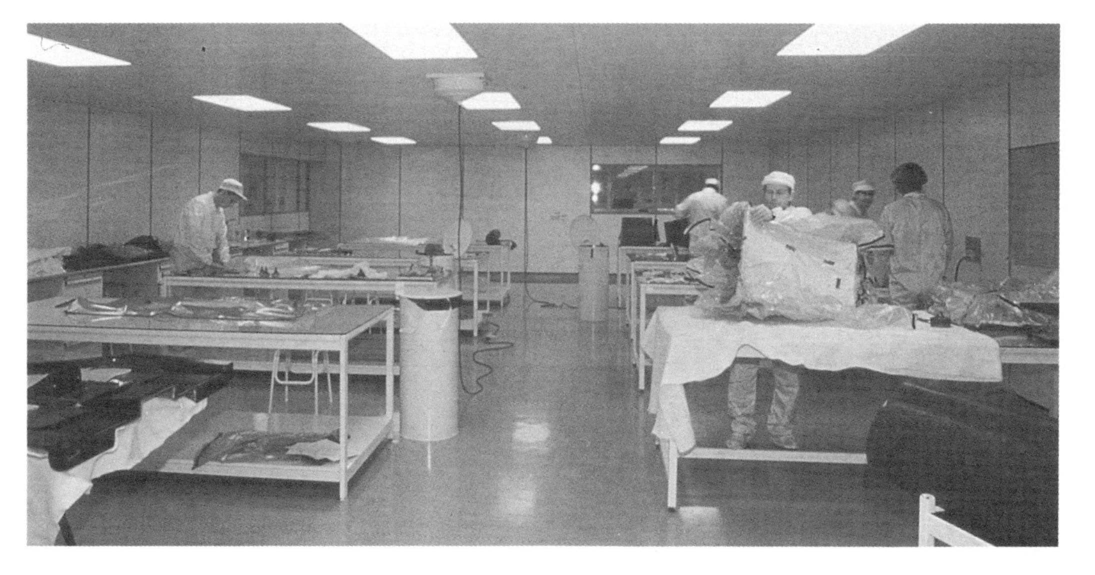

The end result of their deliberations was a very strong carbon fibre/Kevlar monocoque which retained a removable body top to allow access to the engine bay, and the very small shock absorbers which, working in conjunction with torsion bars mounted within the monocoque above the driver's ankles, were positioned beneath the front coaming.

Due to pressure of space caused by the installation requirements of the 12-cylinder engine, the central fuel cell behind the driver was supplemented after the first race by two secondary additional cells cemented to the main structure. These were to become something of a talking point much sooner than perhaps Barnard had expected.

The Berger and Donnelly accidents: how safe can F1 cars be?

Within 18 months during 1989/90, Formula 1 racing would be shocked by two major accidents which would focus even more attention than ever over the way in which contemporary Grand Prix car design was addressed. Moreover, they were accidents which left the F1 fraternity seriously divided over what could and could not reasonably be expected in terms of driver protection from a carbon fibre composite F1 chassis.

The first such accident took place in the 1989 San Marino Grand Prix, when Gerhard Berger crashed his Ferrari 640

The design office at Enstone.

The Benetton Formula race preparation shop.

very heavily during the early stages and was fortunate to escape when the car caught fire. The second occurred in practice for the following year's Spanish Grand Prix at Jerez de la Frontera when Martin Donnelly's Lotus 102 Lamborghini suffered a front suspension failure and ploughed, almost head on, into a guard rail with more horrifying consequences.

Dealing with the Ferrari accident first, it was slightly ironic, bearing in mind what was to befall Berger, to take note of one of the items on the agenda for the meeting of FISA's World Motor Sports Council, scheduled to take place in Paris four days after the Imola race. The item in question was to consider proposals 'to improve the dynamic test procedure (for F1 chassis) in particular by testing the entire monocoque including the fuel tank'.

Berger had been running fifth at the end of the third lap in this particular race, sandwiched between the Williams-Renaults of Riccardo Patrese and Thierry Boutsen, when his car failed to negotiate the flat out Tamburello right-hander immediately after the pits. It slammed off the track into the retaining wall with considerable ferocity.

The ensuing conflagration was the first major fire in an F1 race since Ricardo Paletti perished at Montreal after running his Osella into the back of Didier Pironi's stalled Ferrari 126C2 at the start of the 1982 Canadian Grand Prix. Mercifully, Berger was to be more fortunate.

The crew of an adjacent Alfa Romeo fire tender responded instinctively, arriving at the scene within 15 seconds of Berger hitting the wall and extinguishing the flames within another eight seconds. It was, gratifyingly, a far cry from the hopelessness and lack of understanding which presaged the fiery deaths of Piers Courage, Jo Siffert and Roger Williamson less than 20 years before. F1 had come a long way in terms of both passive and active safety.

The impact in the Berger accident drove the right-hand water radiator, which was angled at about 45-degrees to the centre line of the chassis, back into the fuel cell, which had been topped up prior to the start with pre-cooled Agip.

The Ferrari's carbon fibre monocoque deformed progressively to absorb the enormous impact, but was completely destroyed, its right-hand side ripped open. The driver thankfully survived with superficial injuries, and would race again after missing only a single Grand Prix, but the incident threw into renewed focus the pressures facing F1 designers. In an era when simply to explain that motor racing is dangerous would no longer be accepted as the justification for the occasional fatal accident, designers found themselves walking a tightrope between the demands of track performance and constructional strength which called for increasingly delicate judgement.

Barnard's reflections on the matter give a very clear insight into how seemingly unrelated incidents can conspire to produce a near catastrophe, despite all the care and attention to detail in the world.

'Racing is dangerous and things like this can always happen, but in the wake of this accident it was important for me to try analysing what had actually happened. In building the Ferrari 640, I had tried to make everything as light as possible; the suspension, the wings, everything was lighter than it had been on the turbo car. I managed to save weight on almost every component.

'What also transpired was a rather unfortunate situation because, throughout 1987, we had been running with flexible end plates on the front wings. Since the start of the 1988 season, we replaced them with non-flexible end plates because there was some concern that flexible ones might infringe the regulation prohibiting moveable aerodynamic devices.

'However, I don't think anybody else actually went to a rigid skirt. Most other peoples had cables, allowing the wings themselves to flex, and, when we did the original 639 test car, we designed in flexible construction on the nose wings. But even though they ran many, many kilometres in testing, we still believed that their flexible construction was potentially an

Fabrication work on components destined for contemporary Benetton-Ford F1 chassis are produced in this light and spacious area.

area for performance loss and was possibly causing a handling imbalance.

'As a result, before the 1989 Brazilian Grand Prix, we changed the layout and actually made the front wings stronger, which I thought, after so many miles, obviously meant we would have no problems.

'What I had overlooked was that Imola was the first circuit featuring high kerbs on which we had run the 640. Running over kerbs is something that many drivers obviously do and don't tell you about, but I would also say that it was something I wasn't used to.

'Most of the drivers I had dealt with up to that point, such as Niki Lauda and Alain Prost, knew if they had run over a kerb and tended to mention it. However, I think drivers like Nigel and Gerhard may have got used to running over kerbs quite hard and were not so used to mentioning it.

'In doing that without any small flexibility in the front wing end plate, it eventually over-strained the wing which broke at its inboard edge. It took a few of us at GTO a lot of time to figure out what had happened, because there was a lot of gossip in the press about the fixing points which was completely incorrect. But, if you ask me whether I felt I was to blame for the accident, whenever anything like that happened on one of my cars, yes, ultimately I am to blame.'

As a result of this accident, Barnard went through the Ferrari 640 monocoque design again in painstaking detail and came up with more minor revisions to its construction. Changes to the way in which the carbon fibre composite material was laid up were prescribed, in addition to the fitting of anti-peel bolts, through bolted between the outer and inner monocoque skins, to reduce the likelihood of the secondary fuel tanks becoming detached through impact damage in the event of a similar accident in the future.

It was interesting to note that, at the time, other F1 designers expressed concern that new regulations requiring the pedals to be mounted totally behind the front axle line had, in effect, obliged Barnard to compromise the 640 design by putting the secondary tanks around the side of the driver. Unless the designer chose to extend the wheelbase, which was not generally a preferred solution, it was difficult to find sufficient space to package the entire fuel load behind the driver, as had been the case during the turbo years when the engines had generally been shorter and more compact.

Harvey Postlethwaite, then working for the Tyrrell team which was using 3.5 litre Cosworth DFR V8s, commented:

> 'Fortunately with a Cosworth engine I am not in that situation. However, if someone told me to install a 14-cylinder engine in a car tomorrow, I would have to think long and hard before putting fuel further forward than the driver's backside.'

Donnelly's accident in Spain 18 months later was far more serious. With just eight minutes of the first official Spanish Grand Prix qualifying session remaining, the Lotus slammed head on into the barriers going into the first of a pair of high-speed, right-hand corners immediately behind the paddock. The car hit the slightly curving metal barrier between two supporting posts, the barriers bent back, thus absorbing the impact, and hurled the wreckage, such as it was, straight back on to the circuit.

Such was the ferocity of the impact that the front of the car, ahead of the fuel cell, had totally disintegrated and lay scattered across the track. Donnelly lay in the middle of the circuit, still strapped to his seat, suffering from multiple injuries, and only prompt medical intervention saved his life. Although he would subsequently recover, his F1 career was at an end.

The rear end of the car had been launched several feet into the air and there had been a brief flash fire. The ferocity of the accident understandably shocked the F1 paddock, and many people could not believe that anybody could have sustained such a fearful impact and survived. Lotus technical director Frank Dernie, the man who had designed the type 102:

Benetton Formula has invested heavily in the latest autoclave ovens for the carbon fibre curing process in addition to state-of-the start equipment for its machine shop.

'I'm staggered. I am just amazed that anyone can survive an accident at that speed and at that angle.

'Martin's survival from such a high-speed, head on impact means that FISA's safety measures built into the construction of current Grand Prix cars seem to have saved him from further injury.

'In absorbing the massive energy of the impact, the composite monocoque was fragmented. With any other kind of material used for the monocoque, Martin would probably not have survived.'

Dernie's conclusion was broadly correct. Over two decades, Grand Prix constructional standards had improved beyond belief. But there were still obvious problems to be addressed, and designers were still striving to make even greater improvements even though the optimum packaging of the Grand Prix car design proved to be an ever more complicated affair.

A cautionary note was sounded by Brian O'Rourke, the senior engineer working on composite materials at Williams Grand Prix Engineering. He had joined the Didcot team in 1982, his first project being the structural design of the Metro 6R4 Group B rally car. Thereafter he was responsible for key aspects of the design, manufacture, testing and integrity of all composites employed in the design of the Williams FW10 chassis onwards. In an article in *Automotive Materials* entitled 'Designing for survival', he wrote:

'Aerodynamics are driving the development of the Grand Prix car and very often the chief engineer of a Grand Prix team is an aerodynamicist before, and sometimes rather than, anything else.

'Having produced an optimized aerodynamic configuration and decided on the packaging of key components, the structure tends to be what is left between them. In this situation the structural designer often finds that the performance that he had obtained from his previous design is compromised by the smaller and more complicated shape that he is now presented with.

'In these circumstances there is a tendency for engineers to try and design a material to balance the deficiencies of the structure's geometry. The simple substitution of one type of carbon fibre for another seems an attractive option. In this industry, where engineers' enthusiasm sometimes outweighs their knowledge, the process may appear to be a simple and obvious one; the reality may be very different.

'It comes as a surprise that [in 1991] there is a debate on

how stiff structures need to be. Experience of structural testing, and the application of some simple logic concerning the relative stiffnesses of the suspension components, should lead to the acceptance of an adequate value.

'There are still those who believe that the more stiffness that is available, the better the result will be, and have changed to a high modulus material in spite, or unaware, of the fact that their strength margins may be depleted. An even worse situation occurs when a designer, having achieved the right level of stiffness performance, decided to replace his standard material with a smaller quantity of high modulus fibre to save weight.

'There seems to be an attitude among some engineers in Formula 1 (where beating the regulations is as much a part of the sport as beating the opposition) that as long as the car complies with the letter of the impact regulations then that is as much as is necessary.

'Fundamentally, the traditional racer's approach of trying to cut every available corner to achieve the best result is incompatible with sound structural design unless it is based upon a real knowledge of the subject.'

O'Rourke went on to argue that FISA should compile a catalogue of acceptable carbon fibre material types which achieve 'a defined level of performance'. The sport's governing body fell short of obliging, but for 1991 introduced a series of far more intensive monocoque pressure loading tests to which individual chassis had to conform prior to the start of the new season.

There had been a nosebox impact test requirement ever since 1985, but now a fully representative chassis and nosebox had to be subjected to a 47 KJ impact. The structure and the trolley on which it was mounted had to have a mass of 780 kg, making it the equivalent of a fully loaded, overweight Grand Prix car, and had to be propelled into a vertical barrier at 11 m/s. From the start of 1992 this process of verisimilitude had to be enhanced by the inclusion of a simulated fuel load and a dummy driver in order to check the integrity of the seat belt anchorage points and the bulkhead junctions ahead of the fuel cell.

For the 1991 season, the safety tests were expanded to incorporate the static load applications to the side of the chassis. This was carried out by means of load pads positioned at four different points along the side of the chassis. The chassis had to withstand loadings of 2,000 kg which was increased to 2,550 kg for 1992, while a similar pressure test on the underside of the fuel cell area was also required to 1,000 kg (1991) and 1,250 kg (1992).

CHAPTER 3

AERODYNAMICS

Intensive aerodynamic development has become an increasingly pressing priority within Grand Prix car design since the mid-1970s, and the techniques for investigating this crucial area of performance have evolved to corresponding levels of sophistication to the point that by the start of the 1990s any front-line F1 team worth its salt had exclusive access to its own wind tunnel facilities.

For more than 50 years, aeronautical engineers have employed wind tunnel development to shape the future path of their aircraft designs. The wind tunnel provides the opportunity for the proposed design to be subjected to gale force winds, and by careful monitoring and plotting of the aerodynamic pressures, a clear picture of the prototype shape's aerodynamic potential can be accurately obtained.

In the 1970s, although the Motor Industry Research Association (MIRA) had a full-scale tunnel available, most of the significant racing car development throughout that decade was produced by two quarter-scale facilities, the Donald Campbell tunnel at London's Imperial College and the similar tunnel at Southampton University.

Williams Grand Prix Engineering became the first British-based team to have its own tunnel when in 1980 it acquired the quarter-scale tunnel which had originally been built by Specialised Mouldings, the Huntingdon-based company which specialized in the manufacture of glass-reinforced plastic mouldings used for racing car body panels. This was in service until 1991, when it was succeeded by a highly advanced half-scale tunnel incorporated within a totally new research and development centre at the team's Didcot factory. The supplanted tunnel

The original Williams wind tunnel, which was acquired by the team as long ago as 1980, provided for the aerodynamic testing of quarter-scale models. It was replaced in 1991 by a brand new facility capable of testing half-scale models, and was subsequently purchased by the resurgent Team Lotus.

was then sold off to Team Lotus.

Nowadays a professional racing team would be no more likely to forego access to a wind tunnel than it would do without carbon fibre chassis technology. The McLaren team currently tests with one-third scale models in a tunnel at Teddington owned by BMT, formerly the National Physical Laboratory before it was privatized. Benetton has an exclusive contract with the Royal Military Laboratory at Shrivenham, while Ferrari has its own one-third scale model tunnel, having previously used the Pininfarina facility in Milan.

Much credit for effective early wind tunnel work has been rightly attributed to Shadow designer Tony Southgate during the mid-1970s, when he worked in conjunction with aerodynamicists John Harvey and Peter Bearman in the Imperial College wind tunnel. However, Team Lotus designer Peter Wright was the man who made some of the most significant discoveries which led to the development of Colin Chapman's ground-effect Lotus 78 and 79 cars which were responsible for comprehensively raising the competitive stakes in the second half of that decade.

Whilst working with models of the still secret Lotus 78 in the Imperial College wind tunnel, Wright almost accidentally stumbled over the development which was destined to have enormous implications for F1 car design over the five years that followed.

Wright had been studying the complexities of airflow beneath racing cars ever since the late 1960s when he had worked at BRM before going to Specialised Mouldings – where, ironically, he worked with the wind tunnel with which he would become reacquainted 20 years later, after it had been sold back to the resurgent Team Lotus.

During his spell at BRM, a wing car version of the troublesome P138 chassis had been envisaged, based on Wright's input, but this was shelved during John Surtees's over-influential period as number one driver in 1969 and never subsequently picked up on.

Wright had originally joined BRM in 1966, having obtained a degree from Cambridge where he had specialized in thermo and aerodynamics. Looking back on the P138 'wing car' concept, he reflected, 'In the end I don't think Tony Rudd [then BRM's chief designer] really liked the idea of wings on racing cars in the conventional sense, and although we evolved a prototype side wing design, nothing came of it.'

During the course of Wright's tests to finalize the Lotus 78's basic aerodynamic configuration, which included an assessment as to whether the water radiators could be incorporated within the leading edge of the inverted wing side pods, Wright began to obtain non-repeatable results with the wind tunnel model. What followed was a breakthrough of momentous significance which would open the

The 1991 Jordan 191 was a fine example of an uncomplicated, but aerodynamically very refined carbon fibre composite F1 chassis which was powered by the compact Cosworth-built Ford HB V8 engine. The slim profile of its monocoque was enhanced by the use of a 'mono-shock' front suspension configuration and its outstanding aerodynamic performance centred round a slightly raised nose section and a cleverly conceived arched diffuser panel at the rear.

door to new areas of fundamental understanding of the significance of under-car airflow, establishing a basic yard-stick which would remain valid even after wing section underbody profiles were banned in F1 at the end of the 1982 season.

On closer examination, Wright detected that the models side pods were sagging, and as they moved close to the tunnel floor, so the downforce increased. The Lotus research team explored this phenomenon with a degree of fascination, quickly cutting up some makeshift card-board side panels which extended the models pods right down to the ground. Thus tested, their results indicated that the downforce had doubled, opening the way for skirted ground-effect F1 racers.

Wright remembers that the development of a fully effective sliding skirt system occupied the Lotus R&D team for the best part of the 1977 season, during which Mario Andretti and the late Gunnar Nilsson were achiev-ing excellent results with the steadily evolving type 78. But it was a long hard slog as they worked their way through that development programme.

One of the most fundamental shortcomings with the 78 was that its centre of pressure, the point at which most of the downforce was being generated, was being pro-duced too close to the front of the car. That produced a tendency to snap into an oversteering mode as it turned into the corners, a habit which was balanced out by the use of a large rear wing to compensate. That, in turn, knocked the edge off the car's straight-line speed.

Having developed the Lotus 78 to what seemed a win-ning pitch by the end of the 1977 season, it was absolutely typical that the highly motivated Colin Chapman would galvanize his design team into producing something better for the following year. If the type 77 had started the team along the road back to a restoration of its competitive status, the type 78 sharply focused the thinking towards the benefits offered by ground-effect aerodynamics.

The Lotus 79 carried that philosophy a giant step fur-ther. The design concept called for a brand new car with a slim monocoque, central fuel cell and inboard suspen-sion all round; in short, every facet of its performance subjugated to excellent aerodynamics.

The effect in terms of downforce loadings on the ground-effect Lotus would be quite remarkable. At around 60 mph (97 km/h) the car would generate around 550 lb (250 kg) of downforce, but that would leap to some-where in the region of 3,500 lb (1,588 kg) of downforce at

150 mph (241 km/h). To put it another way, the all-up weight of a stationary car would be multiplied by a factor of four when it was running at high speed, squeezing every ounce of potential from its ground-effect aerodynamics.

A key factor which made the Imperial College results so significant was that the wind tunnel was fitted with a moving ground plane. This refinement had been initiated by Professor John Stollery who, having collaborated as an aerodynamic consultant with the late Donald Campbell on both his record-breaking boat and car projects, was by then involved in detailed research on the behaviour of streamlined shapes operating close to the ground.

The basic benefit of a moving floor, or rolling road, for racing car development, can be summed up as follows. If air is blown along the floor of an ordinary wind tunnel a boundary layer exists, a layer of stationary air attached to the floor. As the distance from the fan increases, the thickness of this layer increases and eventually it may separate from the surface, forming large eddies and a region of turbulent air quite different from the (stationary) air which a real car encounters on a real road.

A moving ground plane is driven by a variable speed electric motor at tunnel air speed, so there is no relative motion between air and the ground, as in the real case. However, it is necessary to suck away any turbulent boundary layer at the front of this moving road which has already built up on the stationary part of the tunnel floor ahead of it.

An ordinary, fixed wind tunnel can give reasonably good results for normal cars with high ground clearances, but they would be totally misleading for an F1 car with negligible ground clearance and extreme dependence on the downforce generated by underbody airflow. The foregoing refers primarily to boundary layers in the tunnel flow, but of course they exist also in the airflow around and below a real car or its model, and the Reynolds Number is a crucial parameter in matching wind tunnel results to full scale aerodynamic forces.

The likelihood and position of boundary layer separation is a function of the Reynolds Number, an aerodynamic formula which owes its name to the experiments with flow of fluids in pipes carried out by Professor Osbourne Reynolds in the early 1880s. He found that there was a critical velocity at which the flow pattern changed from steady to turbulent. Below that point, resistance was proportional to velocity, and above it to the square of velocity.

The boundary layer gradually thickens in the direction of airflow and may separate from the surface altogether when it meets a pressure gradient, for example at the point where the cross-sectional area of the moving body begins to decrease.

When it comes to examining the flow over wings and diffuser panels, for example, which are operating close to the point of stalling, it is important to get the boundary layer conditions correct and the Reynolds Number as close to the full-sized Reynolds Number as possible.

The detailed theories surrounding the boundary layer are extremely complex, as Peter Wright is quick to emphasize:

'Get rid of the boundary layer and everyone could draw what the airflow does quite easily. But it is an extremely complex subject which absolutely dominates the whole question of wind tunnel development. The aerodynamicist, Prandtl, who tried to write the boundary layer equation, once said he was looking forward to dying because then "God could explain to him how the whole thing worked".'

There are also physical limitations to the speed at which the wind tunnel can be operated, a factor which has prompted a trend towards using bigger models over recent years in order to improve the representation.

Put simply, a quarter-scale model run at 50 mph (80 km/h) is the equivalent of a full-scale car running at 12.5 mph (20 km/h). Double the size of the model and it becomes the equivalent of a full-scale car running at 25 mph (40 km/h). Benefits of a larger scale have also come to include more accurate detail on such crucial features as wings where the engineers might be working to tolerances of one thousandth of an inch. Williams Grand Prix Technical Director Patrick Head:

'In order to achieve dynamic similarity – that is to say, an accurate representation in the wind tunnel of how the car's aerodynamics will behave out on the circuit – the closer you get to full-scale conditions, the less the margin of disparity will be. Also, each tunnel obviously has a cross-sectional working area, and depending on the amount of blockage caused by the model within that section, you get errors in the result you end up with. This results from the air speeding up as it approaches the blockage, in this case the model. Obviously, when a car is out on the circuit, there are no such restrictions to the airflow.'

In that connection, Head points out that an F1 car speeding through the tunnel at Monaco, for example, is subject to an adverse aerodynamic effect from the tunnel roof some 15 ft (4.6 m) above it. 'That is why they always tend to oversteer on the right-hander through that tunnel,' he explains.

As an indication of how crucial and detailed today's wind tunnel models have become, by 1991 Williams was employing five specialist model makers with an overall annual budget of around £800,000. Moreover, the team's new wind tunnel, opened in 1992, was a purpose-built facility produced to a rigorous and very demanding set of specifications by a company which had been responsible for constructing one of the latest and most sophisticated tunnels for the Boeing company.

The significant forces which are being measured on F1 wind tunnel models are downforce, drag force and pitching moment. This monitoring is generally carried out by means of a strain gauge balance and is measured electronically. The balance is either mounted on top of the tunnel and the model mounted on struts extending vertically, or the balance is inside the tunnel floor and the model is suspended on a rigid strut joined to the tunnel by the three axis balance measured inside it.

Pressure forces are measured on the surface of the model by little tappings on the surface connected to pressure transducers. These are linked to a computer which can draw maps of the pressure readings and compute the forces involved in the cars pitching and yawing attitude in relation to the ground.

Of course, the overall aerodynamic efficiency of a Grand Prix car is necessarily compromised by the need to attach such accessories as water radiators on to the central fuselage section. This is one of the areas where wind tunnel experimentation has enabled designers to make considerable progress over the years.

As an example, the Lotus 78s and 79s, and indeed the Williams FW07, which superseded the 79 as the definitive contemporary second generation ground-effect F1 design, had top-ducted water radiators. This meant that the hot air exiting the radiators was tumbling out over the upper surface of the side pod, badly disrupting the flow across the rear wing.

Later, side-ducted radiators would become popular in order to overcome this problem, and through ducting, whereby the whole side pod was completely enclosed and the radiator air exited across the rear diffuser panel, would become the most popular option, as pioneered by John Barnard on the Ferrari 640.

However, such configurations could leave engine cooling requirements right on the outer margins, of course, and any mid-season increase of power – or even a particularly hot day – could result in side ducts suddenly being uncovered in order to speed up the exit of that unwanted hot air.

Inevitably, aerodynamic development has been significantly affected over the years by changes in the technical regulations, many of which had their roots firmly embedded within the political considerations of the day. At the start of the 1981 season, sliding skirt ground-effect cars were prohibited by the sport's governing body FISA in what was interpreted by many as nothing more than a decision to slash the edge in chassis technology displayed by the predominantly British-based teams aligned within the Formula One Constructors Association.

With Renault and Ferrari prominent in the switch from 3 litre naturally aspirated to 1.5 litre forced induction engines, any change in chassis regulations which would tip the scales in favour of the engine manufacturer was clearly in their interest. Thus, for 1981 and '82 they were frustrated by the requirement to run fixed aerodynamic side skirts, beneath which a 6 cm ground clearance requirement was made, obliging them to run impossibly hard suspensions in an effort to control the under-car aerodynamics. Throughout these seasons spring rates gradually climbed to the 3,000 lb (1,3661 kg) mark as designers fought to control the fixed skirt aerodynamics, quickly spawning a breed of 200 mph (322 km/h) go-carts from which bruised and battered drivers would regularly emerge to recount horrific tales of close shaves which these difficult-to-drive cars were increasingly unable to avoid.

Over in the Brabham camp, Gordon Murray came up with a ride height lowering arrangement, employing a system of soft air springs which the chassis aerodynamic load compressed as the cars speed built up, dropping the contemporary Brabham BT49C into what amounted to a ground-effect stance. At the time, Brabham was accused of running flexible skirts in conjunction with this system a feature which certainly did not conform with the regulations, but Murray was adamant that his cars conformed fully with the new regulations.

Lotus, meanwhile, had come up with the twin chassis type 86 which had originally been tested in prototype form

One of the Brabham-BMW BT52B turbos being tested in the German car company's wind tunnel, at their Ismaning test centre, as long ago as 1983. The amount of data gathered from such tests was strictly limited due to the fact that this wind tunnel lacked a moving ground plane.

with sliding skirts towards the end of the 1980 season. Wind tunnel tests had convinced Colin Chapman that instead of having separate sliding skirts moving up and down relative to the bodywork, it would be better to spring-mount the body structure atop the wheel uprights, thus transmitting the aerodynamic loadings directly to the suspension and tyres. This system also incorporated a conventionally sprung chassis riding free within the movable aerodynamic body.

This novel idea fulfilled two key requirements at a stroke. In the first place it helped to stabilize the under-car aerodynamics in addition to reducing the physical battering which was an obvious by-product of the ultra-stiff suspension. With FISA in a distinctly combative mood under the stewardship of its controversial President Jean-Marie Balestre, elected to this august office at the start of 1979, it was perhaps inevitable that the twin chassis concept was outlawed before it had even been permitted to take part in a single race. A succession of protests from less enlightened rival teams fuelled Balestre's determination to see the car outlawed, and it was finally written into the history books on the eve of the 1981 British Grand Prix, after six months of gallant effort on the part of the British team.

Having got rid of sliding skirts, the next challenge to the F1 aerodynamicists came at the start of 1983, when sculptured under-body tunnels were prohibited. As from the start of that season, flat bottoms were required for F1 cars from the trailing edge of the front wheels to the leading edge of the rear.

At a stroke, FISA had again wiped out millions of dollars worth of investment and sent the designers back to the drawing boards. John Barnard at McLaren had spent a year directing the design of the new Porsche-made TAG turbo V6 engine to take maximum possible advantage of the ground-effect regulations. Gordon Murray, in turn, had produced a fixed skirt version of the Brabham BT50-BMW, purpose built with a small fuel tank to take full advantage of the in-race refuelling technique which the team had reintroduced the previous summer. In both cases their efforts had been wasted, although Barnard's loss was measured only in terms of possible advantage lost; at least he could still use the same car/engine package, whereas Murray had to start again from scratch. Similarly, the Toleman team's new ground-effect TG183 had to be totally repackaged to conform with the flat bottom regulations, having only made its race debut in the closing stages of the 1982 season.

As Frank Dernie, then aerodynamicist for the Williams team, explained:

'We were now looking at a totally different animal and, suddenly, long side pods tended to produce lift, so the ideal was to keep them short, although we did an enormous amount of experimentation in the wind tunnel to come up with the optimum solution. I tried a big front wing arrangement, sweeping right back through the front suspension assembly, and tried to compliment it with a similar wing type arrangement extending through the rear suspension, which in turn also incorporated the water radiators. But we just couldn't get it to balance; there didn't seem to be any way we could get a worthwhile amount of downforce on the rear of the car, so we eventually gave up on this configuration after about half a dozen tries.'

In revising the Toleman TG183, designer Rory Byrne opted for a full-width nose section into which the water radiators were packaged and the concept also derived additional rear downforce by the positioning of an aerofoil ahead of the rear wing centre line at a point where it could take advantage of the maximum chassis width regulations, complimenting that with a conventional rear wing, positioned slightly lower, behind the rear axle line.

Unfortunately this made the Toleman TG183B enormously pitch sensitive and extremely difficult to set up from circuit to circuit. On the face of it, getting downforce to the rear of the problem doesn't look a particularly difficult problem, the engine and gearbox ensuring that the rearward weight bias of the basic chassis helps matters considerably. However, any aerodynamicist will be quick to demolish that notion, explaining that the front wings use so much of the air that there is precious little energy left for the rear wing.

The optimum function of a wing is achieved when the airflow to and from it, respectively the afflux and tail flow, is not interrupted. Clearly, the rear wing has the biggest problem from the standpoint of afflux, the airflow being broken up and disrupted by the wheels, cockpit fairings and rollover bars. Moreover, with F1 dimensional regulations strictly limiting the height at which rear wings are positioned, there is a restriction on the amount of scope available to a designer for raising it to a level where there is no problem with afflux.

The fact that the engine largely dictated the aerodynamic profile of the car was a matter which preoccupied Gordon Murray when faced with packaging the upright 1.5 litre BMW single turbo engine in the early 1980s, this unit in particular defying all his efforts to achieve a decent airflow over the 1985 Brabham BT54's rear wing. He reflected on this in 1986:

'For some time it had occurred to me that we were forgetting some fundamental principles of racing car design, one of which is that the centre of gravity should be as low as conceivably possible. At the time we were getting rather frustrated with that vertical BMW engine since we could spend a week working away in the wind tunnel and not even find half a second a lap. We started the 1985 season knowing that there wasn't much we could do about it, but Nelson Piquet had been pressing us hard for some time to do a low-line car, and in 1986 we decided to have a go.'

The result of these design deliberations was the low-line Brabham BT55, powered by specially made BMW engines canted over at an angle of 72-degrees. Unfortunately, although the BT55 developed up to 30 per cent more downforce than its immediate predecessor, somewhere along the line it also picked up an excessive amount of aerodynamic drag. Add to that a shortage of power caused by inadequate oil scavenging properties displayed by the angled BMW engine, and the whole package added up to a disaster.

Nevertheless, no F1 designer would ever underestimate the benefit to be gained by reducing the frontal area of the car, and Nelson Piquet in particular remained preoccupied by the need to sit as low as possible in the cockpit, a trait he took with him when he switched to the Williams-Honda squad at the start of 1986. This had its origin way back in his F3 days when he used to tuck down another inch or so in his Ralt on the straight at Silverstone in order to pick up another couple of hundred revs; a worthwhile consideration in this closely matched junior formula.

Nelson badgered Patrick Head to produce him a personalized, low-line version of the already compact Williams FW11B, anxious to capitalize on any advantage he could gain over his team-mate Nigel Mansell. The Englishman, incidentally, quickly became equally concerned about as low a driving position when he discovered that by driving without a seat in his FW11B, he lowered his position in the car by half an inch which translated into an additional 50 lb of downforce!

Rory Byrne later exploited the rules on under-body aerodynamics to the absolute maximum in 1986 when the Benetton B186 featured a distinct V-shaped profile beneath the nose section, deflecting airflow from beneath the main mass of the car while at the same time maximizing the potential for downforce at the front end. This was soon adopted by many rival teams as the quest for as much maximum downforce from flat-bottomed cars was pursued with sustained vigour.

Running close to the ground spawned a veritable flood of aerodynamic information which was of no relevance to any passenger car application, a reality which was concerning many teams by the start of the 1990s as increasing proportions of annual budgets were now regularly expended in pursuing this specialized area.

Commented Lola Cars founder and Managing Director Eric Broadley:

> 'My feeling is that the F1 aerodynamic rules are all wrong. The flat bottom rule is a travesty of engineering sense, because it is totally wrong from a functional, logical point of view. You're running the car as low as you possibly can all the time, and it rubs on the ground – you can't afford to have much suspension movement because of the aerodynamics. The flat bottom rule must be changed.'

Broadley believes that the Indy car regulations, which allow a degree of ground-effect performance, make much more sense. They will be adopted for F1 in 1995.

'It is a ground-effect design still, but it is very limited because the gap between the car and the ground is 2 in, so the leakage is very high, so you don't get a lot of downforce. But it is physically, aerodynamically, a good shape. It's curved underneath so that we can get some stability from it; it's just right, and we run those cars with quite soft suspension.

'On road circuits we run them quite soft with about 1.5 in of wheel movement at the front and maybe 3 in at the rear. In an F1 application, 10 mm is regarded as a lot and some cars, such as the Ferrari in 1992, ran about 5 mm movement.

'The Indy car still bangs its bottom on the ground, but with the practical advantage that you're only hitting the narrow middle bit under the monocoque, which you can make out of something solid and replaceable. You don't destroy the whole floor as you do on an F1 car, and that's one of the big problems. You're destroying these flat bottoms made of carbon fibre. It's expensive, it's ridiculous and, engineering-wise, it's nasty. We should change now to a carefully controlled underbody shape for F1 design.'

There were also a raft of potential problems facing the Grand Prix designers, notably the need to be absolutely certain that the aerodynamic information obtained from the many wind tunnels now in use was absolutely accurate and valid.

As an example of this, in 1989 the Leyton House team had been hamstrung by seemingly endless difficulties

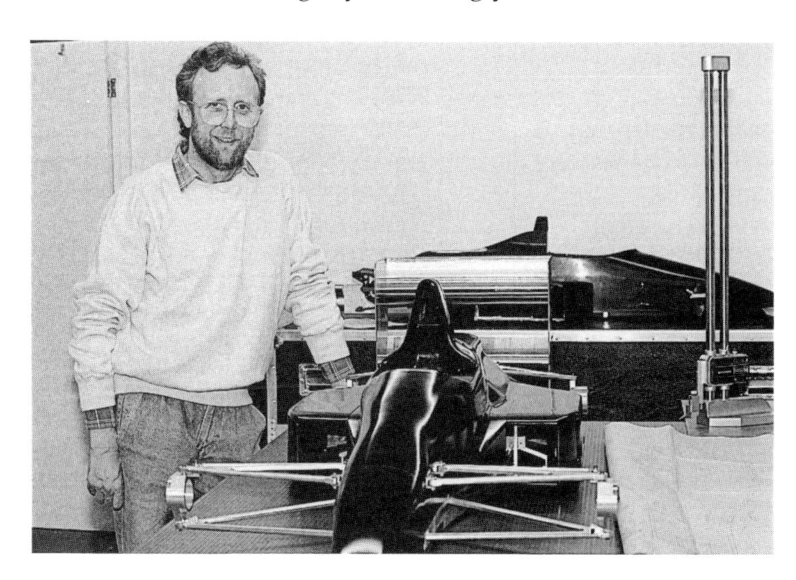

Designer Alan Jenkins poses with wind tunnel models of the 1989 car designed for the now defunct Onyx team. The accuracy of detail required of these quarter-scale models became absolutely central for efficient aerodynamic development as the pace of F1 design technology increased throughout the 1980s.

Tyrrell designer Harvey Postlethwaite (right) and aerodynamicist Jean-Claude Migeot pose alongside the Cosworth DFR-designed Tyrrell 019. The under-cut nose section represented a significant aerodynamic breakthrough which was subsequently imitated by all rival teams.

Right *The need to produce a balanced overall technical package in F1 was illustrated when the Mugen V10-engined Tyrrell 020 failed to live up to expectations raised by the Cosworth-engined 019. The Japanese V10 was so heavy, and needed such an increased fuel load, that in relative terms it proved no more successful than its predecessor.*

with the aerodynamically sensitive CG891 chassis, but the team's Technical Director Adrian Newey started 1990 confident that he had got to the bottom of the car's extreme aerodynamic sensitivity. Unfortunately, the new CG901 chassis proved even worse than before, neither of the team's entries even managing to qualify on the bumpy Autodromo Hermanos Rodriguez at Mexico City.

On the face of it, these problems did nothing for Newey's reputation, but the fact that Williams Technical Director Patrick Head recruited him in the summer of 1990 to take over the key role as chief of aerodynamic development for the Didcot-based team endorsed the reality that his reputation had not been affected by this problem.

In fact, before leaving Leyton House, Newey bequeathed the team a new undertray and diffuser design which helped drivers Ivan Capelli and Mauricio Gugelmin to vault up to the front of the field in the French Grand Prix on the ultra-smooth Paul Ricard circuit. Capelli's performance in this race, where he led commandingly and only relinquished the lead to Alain Prost's Ferrari 642 in the closing stages, provided a graphic example of how slight aerodynamic variations in F1 could make the difference between a non-qualifying also-ran and a potential winner. Sadly, the downside to this situation was that the drivers often got the blame and were forced to carry the can.

A head-on shot of the 1992 Ilmor-engined Tyrrell 020B shows the distinctive anhedral nose wings to good effect.

However, without question, the most ambitious and innovative chassis design of 1990 sprang from Harvey Postlethwaite's design team at Tyrrell Racing, in the form of the type 019 which supplanted the promising 018 at the San Marino Grand Prix. Working in conjunction with aerodynamicist Jean-Claude Migeot, Postlethwaite revived a concept with which he had toyed whilst at Ferrari, namely an upswept frontal profile to the underside of the monocoque working in conjunction with anhedral

*Adrian Newey, who would later go on to produce the aerodynamics for the 1992 Championship-winning Williams FW14B, gained a reputation for attention to detail and neat packaging with his Leyton House/March designs in 1988 and '89. The March 881 carbon fibre composite monocoque (***right***) was one of the slimmest in the business and this shot shows off the elegant simplicity of a contemporary F1 double-wishbone/push-rod front suspension configuration, designed to produce minimum aerodynamic turbulence ahead of the radiator ducting. The profile of the front wing end plate (***right***) maximizes the potential for brake cooling, while at the rear end of the car (***below***) and (***top right***) the bodywork is waisted tightly in around the Judd V8 engine and the packaging of the twin coil spring/dampers parallel to the car's centre line also contributes to cleaner airflow across the diffuser panels. Newey also pioneered the distinctive arched diffuser (***far right***) through which the exhaust pipes also exited for added downforce.*

front wings running close to the ground.

Postlethwaite and Migeot had originally planned a similar configuration for the Italian team. But one which went a step further as they envisaged incorporating the water

radiator beneath the car, but the more straightforward concept they evolved for the Tyrrell produced significantly more downforce without the need to use a large rear wing, and the straight-line performance of the admittedly very light and compact new 019 raised more than a few eyebrows throughout the season.

In fact, this concept derived directly from the indentation in the floor of the monocoque of Adrian Newey's March 881 ahead of its mandatory flat-bottom area. This step successfully regrouped the airflow prior to its rush across the flat bottom, ensuring a high-energy feed as well as shortening the effective length of the underwing, further reducing pitch sensitivity.

When he joined Williams in the summer of 1990, Adrian Newey had only five days in the Southampton wind tunnel in which to evolve the aerodynamic profile of the new Williams FW14. He went for an understated version of the Tyrrell high nose concept, and this, supplemented by all the other technical refinements incorporated into the 1992 Championship winning Williams FW14B, would irrevocably confirm a design trend which every other F1 constructor would pursue in due course.

However, by the start of 1993, Patrick Head was prepared to concede his belief that Grand Prix racing could benefit from long-term technical changes over the coming years, embracing major restrictions on aerodynamic downforce as a means of slowing the cars and producing closer racing. However, while he supported carefully structured, sensible proposals for long-term change, he counselled against sudden changes which had not been properly thought through.

'I think we should look closely at the long-term regulations, but this is a package which has to be discussed and considered without any rush. Generally, I think the way in which we go about producing changes in F1 is incredibly amateur relative to the overall professionalism of the sport.'

Head favours major reductions in downforce to produce better racing. He insisted:

'It would not be any more dangerous. F1 cars go round corners at Hockenheim with 50 per cent of the downforce we use at Monaco. If we ran Hockenheim downforce settings at Monaco, I think the cars would be much more entertaining to watch.'

CHAPTER 4

FORMULA 1 ENGINE DESIGN

The onset of the 3.5 litre naturally aspirated formula 1 regulations in 1989 promised more evenly matched engine performance than had been the case for much of the turbo era, so more than ever the demands facing an engine design team could not be considered in isolation from those which shaped the overall F1 package.

On more than one occasion, Ferrari's Technical Director John Barnard had observed wryly:

'In recent years, it has been my experience that many engine makers have produced their designs without any thought as to how they might fit in the chassis.

'When we were considering a turbocharged engine for the McLarens back in the early 1980s, we eventually decided to commission Porsche to make us one to our precise specification, as no existing unit could be adapted to fit our exact requirements.'

Barnard's exacting approach to the development of the Porsche-made, 80-degree twin turbo TAG PPO1 engine was to represent a major watershed in technical thinking, because this was the first purpose-built Grand Prix engine of its era – apart from the Ferrari V6, which owed nothing in terms of configuration or components to anything that had gone before.

All its rival F1 turbo engines had their roots in other applications. The similar V6 engines from Renault and Honda had been spawned from early F2 units; BMW's four-cylinder 12/13 was based on a production block, and Brian Hart's praiseworthy 415T four-cylinder owed much of its development to the Hart 420R 2 litre four-cylinder

Right and far right
The remarkable Honda RA168E 1.5 litre V6 turbocharged engine won the Constructors' Championship for Williams in its 4 bar/195 litre trim in 1987 and then continued to dominate with McLaren the following year, when it won 15 out of 16 races despite being further restricted to 2.5 bar boost pressure and a fuel maximum of 150 litres.

which had powered the Toleman team to victory in the 1980 European F2 Championship.

The turbo era developed predictably into an exercise in developing prodigious power outputs with massive doses of boost pressure, which inevitably ensured short engine life. After being stringently rationed to 150 litres of fuel and a meagre 2.5 bar boost pressure in 1988 – a challenge met magnificently by Honda, who still managed to propel McLaren to a record 15 wins out of 16 races – the turbo era closed for good to be replaced with a requirement for 3.5 litre naturally aspirated engines.

There were many who naively believed that reintroducing non-turbo rules would mean a return to the 'Cosworth formula' of the 1970s when Ferrari's presence as the lone constructor on the Grand Prix scene would be supplemented by a rash of special builders, all using identical Cosworth off-the-shelf DFV V8 power units. But the flood tide of turbo technology had raised technical standards in F1 to levels hitherto undreamt of. The big battalions of the world's motor manufacturers were now involved and the stakes would remain as high as ever, the pace being forced dramatically by Honda.

The rules had been framed to limit the number of permissible cylinders to 12. In truth, of course, it was hardly necessary for such a regulation to be committed to the F1 statute book. Ever since BRM's hopelessly over-complex and heavy H-16 engine had proved such a lamentable failure in the mid-1960s, engine designers had shied away

from such ambitious schemes, effectively realizing that the choices lay between a V8 and a V12. Or so it seemed.

Cosworth's superbly efficient, light and reliable DFV had been the benchmark by which the former had been judged. Yet with piston area and engine speed obviously

The Cosworth DFR 3.5 litre V8 was introduced for use by the Benetton team in 1988 as the first generation Ford F1 engine for the new naturally aspirated post-turbo F1 era. It displayed the same characteristics of reliability and serviceability that characterized its 3 litre DFV forebear and endured in service through to the 1990 season.

the cornerstone on which to build increased power output, the theoretical advantages offered by a V12 were considerable. However, the main problems involved in such a layout had traditionally been excessive fuel consumption and an inherent weight handicap when compared with the V8s. More power was necessary to compensate for the higher all-up start line weight.

As a result, by the early months of 1987 both Renault and Honda were working away at splitting the difference, designers Bernard Dudot and Osamu Goto producing V10 designs for their respective companies. In Honda's case, it proved to be the perfect balance. The high-revving Japanese 72-degree Honda V10 consistently developed in excess of 670 bhp during the first season of the 3.5 litre regulations in 1989. By the end of the following year the Honda V10 topped the 700 bhp mark, at which point it opted for the V12 route, convinced that the greater piston area was a major priority.

Nevertheless, the need for the complete technical package to come out as close as possible to the 505 kg minimum weight limit led many people to conclude that the V12 route imposed an unacceptable straitjacket on chassis designers, who might find themselves precluded from

Honda's RA109E V10-cylinder engine was introduced at the start of the 1989 season and progressively developed to the point where it produced around 685 bhp at 13,000 rpm by the end of the year.

pursuing various other innovative avenues by the overriding weight consideration.

Weight was always the bottom line. At the start of 1989, when the design of the first McLaren MP4/5 chassis was well advanced, Honda, who had completed all the development work on the new V10 using belt-driver camshafts, raised the question of changing to a gear drive mechanism. McLaren was asked to consider whether it could 'come to terms with the weight penalty': F1 'techspeak' for 'Can you save a corresponding amount from the chassis weight?' – and it was quickly decided that the benefits of more accurate valve timing outweighed the extra weight from a long-term viewpoint.

Renault, meanwhile, continued the development of its 67-degree V10, also initially with belt-driven camshafts, and this made its race debut with Williams in 1989. However, the relative lack of success achieved by the Williams-Renault partnership in its early days reflected, at least in part, some of the problems which John Barnard had highlighted back in 1984.

After the French company's withdrawal from the turbo F1 arena at the end of 1986, Renault Sport chief engineer Bernard Dudot was given permission to continue a low-

In 1990 the 72-degree Honda V10 was retained for the exclusive use of the McLaren team, although now curiously dubbed RA100E. By the end of this season its power output was nudging the 700 bhp mark.

key development programme for the new 3.5 litre regulations, just in case a return to Grand Prix racing should be put back on the agenda at some moment in the future: as indeed it was.

Consequently, although Dudot's V10 configuration was to prove absolutely the correct choice, the original RS1 version was not tailored to suit any particular chassis, a fact which inevitably caused the first Williams-Renault FW12Cs and FW13s to be something of an overall design compromise. For 1990, the V10 was revised to sit lower in the chassis, thanks to the incorporation of different engine mounts, a revised cooling system and a lower crankshaft line – which in turn called for significant alterations to the Williams transverse, six-speed gearbox.

However, in the battle for revs Renault retained one of the strongest technical cards of all, a system of pneumatic valve actuation which did away with conventional valve springs.

It was back in 1986 that Renault Sport produced the first Grand Prix engine to feature this refinement. At that time, the *Regie*'s sole obligation in the sport's most senior category was to supply its prodigiously powerful twin turbo 1.5 litre V6s to Team Lotus for the JPS-backed cars, then driven by Ayrton Senna and Johnny Dumfries.

After eight years in F1 with its own factory team, Renault Sport in its own right had withdrawn from the Championship at the end of the previous year, clearly realizing that an alliance between engine manufacturer and small specialist racing team offered a more promising conduit to Grand Prix success than trying to do everything oneself. This left only Ferrari building both F1 cars and engines, and while the Italian team had challenged strongly for the 1985 World Championship, it was clear that McLaren-TAG/Porsche, Williams-Honda and Lotus-Renault had become consistently the strongest players in the business.

Prior to the 1986 season, the Renault engines furnished to Lotus were supplied by the Bourges-based Meccachrome company, effectively a satellite sub-contractor of the main Renault Sport R&D department at Viry-Chatillon. These facilities were shared by Ligier and Tyrrell, as fellow customer teams, but with the Renault works team's withdrawal, Lotus inherited what amounted to works-supported status.

The development and packaging of the 1.5 litre turbocharged Grand Prix engine during the 1980s became more complex every year. Regulation changes seeking to stem spiralling power outputs in the interests of reducing speeds were introduced with monotonous regularity.

Specifically for 1986, although there was no restriction on turbo boost pressure, the reduction in fuel tank capacity from 220 to 195 litres – originally scheduled for the previous year, but deferred – was finally implemented. Every element of the engine's performance had to be maximized for optimum efficiency in order to eke out a full race distance on the new allowance.

It was against this technical backdrop that Dudot and the Renault Sport development team made a quantum leap forward with the introduction of their pneumatically activated valve closing system. Eliminating conventional valve springs was a step which would allow the engine to rev faster, by reducing reciprocating mass and eliminating the over-stressing of the springs due to surge, vibration and valve float, considerations which had dogged conventional racing valve gear systems over the years.

Interestingly, the whole project might have remained totally confidential had it not been for the departure of Renault Sport engineer Jean-Jacques His during the winter following the closure of the French factory team. He had been closely involved with the development of the DP – *Distribution Pneumatique* – and his switch to the Ferrari team meant that the cat was out of the bag.

In conjunction with a totally new engine management system, the DP valve-return system offered several potential benefits for the new Renault EF15 V6. Most significantly, it eliminated the mass and complexity of 48 tiny valve springs, two per valve. Replacing the individual spring was a tiny piston-cylinder arrangement, the piston itself being attached to the valve stem to compress the nitrogen which now formed a gas spring that returned and closed the valve. The whole system was pressurized to between 1.2 and 1.8 bar.

To the outsider, the only give-away was a distinctive half litre aluminium container within the Renault engine vee. Some wilder elements of the media got firmly on the wrong track when it came to identifying this component, suspecting that it perhaps contained some illegal power-boosting additive which was being surreptitiously introduced to the fuel. In fact, it contained just sufficient nitrogen to sustain the system through a 200 mile Grand Prix.

The effects of the pneumatic valve-gear on the Lotus 98T's performance were immediate. The maximum engine speed was increased at a stroke from 11,000 to 12,500 rpm, providing the drivers with a wider power band which in turn made the task of gearing the car for different circuits somewhat easier. Dudot also admitted that another benefit accrued from the fact that the drivers could now over-rev the engines without any dire mechanical consequences. During the course of the season, some units were momentarily over-revved to 13,500 without any adverse effects.

Renault, of course, withdrew from F1 at the end of 1986, after a season which had seen Senna put the pneumatic V6 turbo on pole position nine times, although winning only two of the 16 races. It was fair to conclude, with the matchless benefit of hindsight, that the potential of this French engine was severely under-estimated, for although Senna was hell-bent on having a Honda V6 turbo for 1987, the resultant Lotus-Honda package proved, in relative terms, scarcely more successful than the Lotus-Renault.

From the outset of the V10 development, Dudot retained a virtually identical pneumatic valve closing system, mindful that increasing revs would be a crucial key to steadily increasing the power output of a naturally aspirated F1 engine. Thus by the end of 1991, Dudot's RS3B V10 was developing around 770 bhp at 14,200 rpm, and at the height of Nigel Mansell's World Championship onslaught the following year, the power had been boosted

Porsche's brief foray into the world of 3.5 litre F1 racing engines served as a salutary lesson to those contemplating this high-tech arena. Supplied to the Footwork team in 1991, this 80-degree unit, fully dressed with clutch, flywheel and other ancillaries, tipped the scales at over 418 lb (190 kg) at a time when the Cosworth-Ford HB weighed in at just under 300 lb (136 kg). This was an impossible weight handicap to surmount even without the oil scavenge problems which also afflicted the project and which was shelved amidst much acrimony before the end of the season.

beyond the 800 bhp mark with the engine now running close to 15,000 rpm.

Meanwhile, at the V8 end of the scale, the Cosworth-built Ford HB engine, which supplanted the DFV-derived DFR V8 at the start of the 1989 season, was running to 13,500 rpm by the middle of the 1981 season, at which point its power output was approaching a remarkable 670 bhp. This, allied to the advantage of low all-up weight and frugal fuel consumption, meant that it was comfortably establishing a reputation as best of the rest behind the trio of front-line multi-cylinder runners, Honda, Ferrari and Renault.

Cosworth had also been actively considering the development of its own pneumatic valvegear system. After initially considering the system for the 3 litre naturally aspirated DFY V8, the Northampton concern shelved the programme when Ford switched to the 120-degree turbo of 1986/87. Cosworth concluded at the time that it could control the tiny valves quite adequately with normal springs.

'Then we got back to 3.5 litres with the DFR in 1988', explains Dick Scammell, Cosworth's racing manager, 'and we were back into a situation where we had to deal with even bigger valves and we really needed to look at the system again.' One of the biggest problems Cosworth encountered from the outset was controlling leakage from the system.

Meanwhile, the progress made by Cosworth with the HB V8 was putting pressure on the opposition to raise the standard of their game. As the highly respected and experienced F1 race engine designer Brian Hart explained:

> 'By a process of logical extrapolation, taking into account all the factors involved, with the HB V8 running at 13,500 rpm, a V10 would have to be operating at around 14,200 rpm and a V12 approaching 15,000 rpm if their potential is to be maximized. And that poses obvious problems, because not many people have managed that before.'

Brian's predictions for the Cosworth HB's potential proved almost exactly on target. By the start of the 1992 Grand Prix season, the Series VI HB V8s powering the Benetton B192s were running to 13,500 rpm, equipped with pneumatic valvegear and dynamic crankshaft balancers, enabling Ford to lay claim not only to having the fastest-running V8 engine ever, but also to having boosted output to around 740 bhp, a figure which would be topped again by the Series VII V8 in the middle of that season. As a result, Benetton was now drawing level with McLaren-Honda, challenging to become the second force in F1 behind the Williams-Renault.

Of course, very little firm data as regards the bore and stroke dimensions of contemporary F1 engines have been made available, leaving the technical observer facing a series of tantalizing conundrums. Without these two dimensions, accurate assessments of relative potential performance is almost impossible. However, there are those in the Grand Prix business capable of making calculated guesses, most notable of whom must be Brian Hart. When his estimate of the bore/stroke figures for the 1989 72-degree Honda RA100E V10 was published, one could watch the colour drain from the faces of the Japanese teams engineers...

The relationship between piston and valve area is central to the different thinking behind these F1 engine configurations, very high engine speeds requiring very large valve areas, while piston area is kept as small as the required valves permit.

In comparative terms, it has also been estimated that the Ford HB V8s, for example, were in 1992 giving away

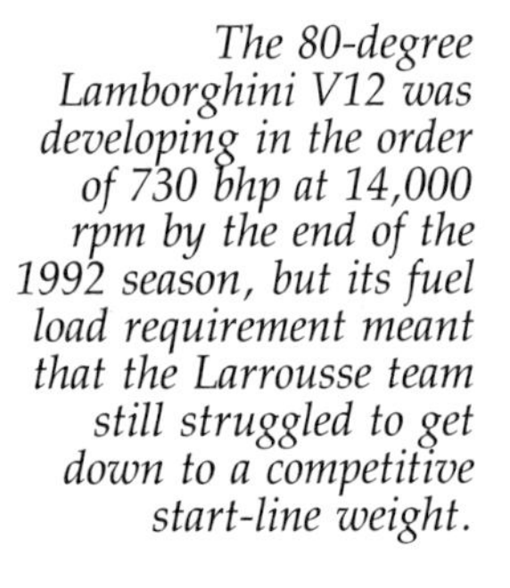

The 80-degree Lamborghini V12 was developing in the order of 730 bhp at 14,000 rpm by the end of the 1992 season, but its fuel load requirement meant that the Larrousse team still struggled to get down to a competitive start-line weight.

some 18 per cent in piston area to Renault and Judd GV V10s, and 22 per cent in piston area to the Honda RA121E/B, Lamborghini 3512, Yamaha OX99 and Ferrari E1A/92 V12s.

To a lesser extent, Ilmor also seems to have sacrificed bore area on its superb little 2175A V10, which is 2.5 mm (1 in) shorter than the V8 Ford HB. Despite this compactness, the Ilmor V10 – which is to be used by the Mercedes-Benz-supported Sauber F1 team in 1993 – has apparently some 7 per cent more bore area than the Ford V8. On the other hand, with about the same bore size as a representative 12-cylinder engine, the Ilmor V10 gives away some 17 per cent in terms of bore to the Honda, Ferrari and Yamaha V12s.

After two World Championship-winning years with its V10 engines powering the works McLarens in 1989 and '90, Honda proved unable to resist the lure of a V12. Unfortunately, this preoccupation with valve area would upset the delicately balanced McLaren-Honda performance envelope and lead to a significant drop off in achievement for the Anglo-Japanese partnership in 1992.

The first worrying signals could be seen at a pre-season test at the start of 1991. Ayrton Senna returned from a

The superbly compact Ilmor type 2175A 72-degree V10 engine arrived on the scene at the start of the 1991 season, and despite comparative lack of finance, proved itself to be an efficient and reliable unit. In 1992, when the Northampton-based company built a new V10 specially for the Mercedes-backed Sauber team, Ilmor's F1 prospects looked set to strengthen dramatically.

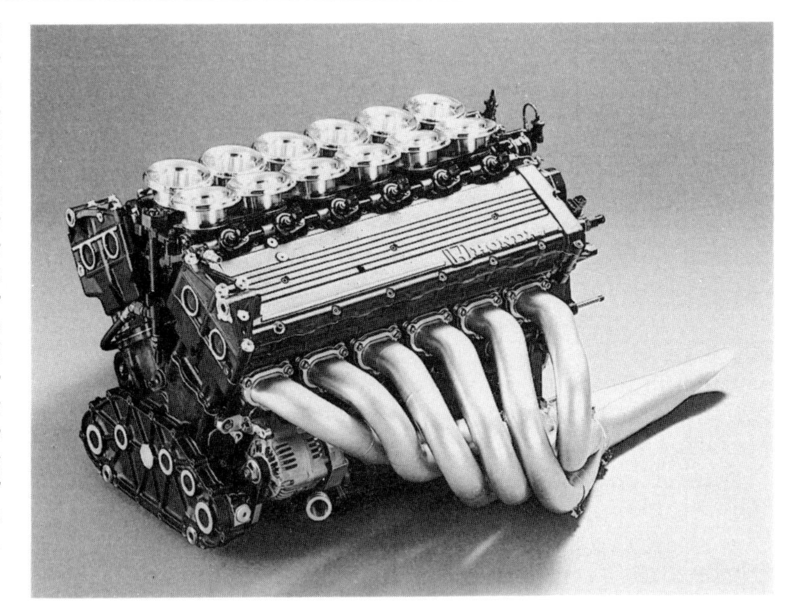

The Honda RA122E/B engine had a wider (75-degree) vee angle than its immediate predecessor, but although the V12 benefitted from pneumatic valve actuation and variable length induction trumpets, its power output was disappointing and it had a voracious appetite for fuel. It was largely responsible for the McLaren MP4/7A being such a disappointing car by the team's high standards.

sabbatical in his native Brazil to find that very little progress had been made with the Honda RA121E V12. The Brazilian would go on to retain the World Championship, but this was largely due to McLaren's efforts in shaving weight from the MP4/6 chassis throughout the season, just turning the Williams-Renault V10 tide which was running strongly against them in the middle of the year.

To suggest that it had been a mistake to opt for the V12 over the V10 was a view increasingly whispered within the McLaren-Honda ranks. For 1992, prospects took a fur-

The 60-degree Honda RA121E V12 engine supplanted the V10s for McLaren's use in time for the start of the 1991 season, reputedly developing 780 bhp in 14,800 rpm at the zenith of its development. However, Ayrton Senna was sceptical about its performance early in the season; rightly so, as the unit suffered excessive internal frictional losses, and consequent heat retention, which compromised its performance. In addition, it suffered from a spate of main-bearing failures which were caused by unbalanced oil distribution to the crankshaft.

ther tumble with the advent of the brand new 75-degree Honda RA122E/B engine powering the McLaren MP4/7A. The result was that McLaren slipped from the high wire. Although winning five races, the season was adjudged an acute disappointment by their standards. It was a cruel, classic lesson in how crucially interrelated are all the ingredients which combine to contribute to the overall performance of today's highly complex F1 car.

The new Honda engine, with its vee angle set 15-degrees wider than its immediate predecessor, was intended to offer more power through higher revs thanks to the use of pneumatic valve gear. The new V12 also featured variable-length induction trumpets, first seen on the 1991 engine, as well as a new inlet port design intended to capitalize on Honda's 'quick burn' technology derived from the company's work in meeting exhaust emission control regulations for road cars.

The stark disappointment, however, was that the new engine failed to produce the expected performance. To start with, it was late coming on the scene. It was too heavy, handicapped by serious internal frictional losses and, partly as a result, was extremely heavy on fuel.

At Silverstone, just prior to the start of the British Grand Prix, the McLaren MP4/7As were the only cars to be topped up with fuel on the starting grid, and Senna's struggle with Martin Brundle's Ford V8-engined Benetton can be put into sharper perspective by the fact that the Honda-engined cars started the race with over 220 litres of fuel on board, compared to around 185 litres for the Cosworth car.

That represented an extra weight handicap in fuel alone – which can be doubled when one takes into account the relative weights of the V8 and V12 engine. With that in mind, it was hardly surprising that Senna had so much trouble competing. Not until the very end of the year had Honda got the frictional losses fully under control to the point where the MP4/7A was fully competitive with the Williams-Renault FW14B.

From the outset of the 3.5 litre F1 technical regulations, Ferrari had followed its traditional route and concentrated on a V12 configuration. The design of the 65-degree V12 was dictated by John Barnard during his first stint with the Italian team (1986-89) on much the same basis as he had instructed Porsche on the V6 TAG Turbo project. As far as the Ferrari unit was concerned, he originally thought in terms of a 60-degree V12, but opened it out by another 5-degrees in order to provide more room to pack-

Putting the finishing touches to one of the Cosworth-made Ford HB engines at the firm's Northampton headquarters. For 1993 these engines were supplied with works support to the Benetton team with customer engines going to McLaren, Lotus and Minardi.

age ancillaries within the vee.

Ferrari resolutely stayed with the V12 configuration ever since the start of 1989, only able realistically to get on terms with Honda in the closing races of 1990 when AGIP appeared briefly to produce a performance edge over rival products from Shell (Honda) and Elf (Renault) in the fuel race.

Thereafter, Maranello's engine department seemed to freeze. By the start of the 1992 season, the V12's performance had become positively arthritic. Blighted in the early stages of the season by oil system problems which caused a number of bottom end failures, the engines improved in reliability, but not performance, giving away as much as 100 bhp in real terms to Honda and Renault. Combine that with it having none of the benefits of a small, light fuel load which accrued to the V8 competitors, and Maranello's prospects for 1993 looked pretty dire by the start of that year.

In many ways, 1992 was a pivotal season for Grand Prix engine design. Yamaha, who took the V12 route in partnership with the Jordan team, almost sank without trace, the engine producing insufficient power despite an insatiable appetite for fuel.

For 1993, Yamaha opted for a partnership with Judd,

basing their next F1 power unit on a joint development of that company's type GV V10, while Brian Hart also opted for this configuration on his company's return to the front-line F1 action with the type 1035 engine, ironically enough supplied to Jordan as a replacement for the unmourned Yamaha V12!

By the start of 1993, it was becoming clear that the V12 was a dead duck as far as F1 was concerned, only Ferrari and Lamborghini continuing to pursue this fuel-hungry route. Ford, who had grandly announced plans for a Cosworth-built V12 during the summer of 1991, now seemed content to let this simmer away on the back burner. It may have been capable of running to 16,000 rpm, but even that wasn't sufficient to produce a significant edge over the ever-improving HB V8 derivations, and at the time of writing, it seems unlikely that it will ever race.

Management systems

Of course, fuel technology and the efficiency of the complex engine management systems, which have become an essential accessory over the past few years, have assumed

The engine that beat the Hondas. Bernard Dudot's 67-degree Renault V10 first made its appearance on the scene in 1989. By the start of the 1992 season, both the RS3 and RS4 derivatives were poised to hoist the Williams team into an unparalleled period of F1 domination.

an absolutely crucial level of significance. Matching ignition and fuel mixture across a wide range of revs to ensure optimum operational efficiency and fuel consumption has called for the development of highly sophisticated engine management computers, many of which have produced worthwhile benefits for the manufacturers involved.

In this connection, it should be pointed out that the Honda NSX direct ignition system, PGM-F1 fuel injection, double overhead camshaft, four valves per cylinder configuration and titanium rods are directly attributable to the expertise gained by Honda from its F1 involvement. On the same basis, the Ford EEC-IV management system employed on the Cosworth-Ford HB engine is directly related to the system evolved by Ford electronics for use across a wide range of their road cars.

Engine management systems were introduced into F1 during the 1980s, effectively succeeding the characteristic 1970s configuration when capacitor discharge ignition systems working in conjunction with mechanical fuel injection, represented state of the art technology. With these systems, fuel delivery increased in response to increased load, and rather than constant flow, injection was timed to correspond with valve opening.

Increasingly, these systems tended to involve compromise settings, as there was no existing system which

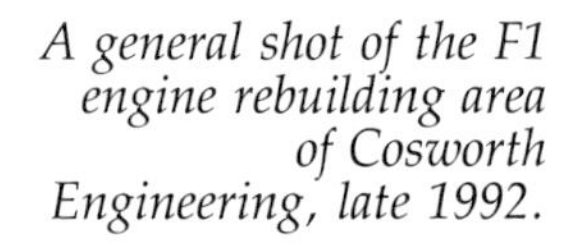

A general shot of the F1 engine rebuilding area of Cosworth Engineering, late 1992.

could respond quickly enough to continually changing conditions of engine speed and load. The same carefully timed fuel measure would be injected for a given engine load, regardless of the revs. Thus, it was always necessary to strike an operating balance between a setting which was sufficiently rich to avoid internal over-heating, but which produced a down side in high fuel consumption for a given power output, and generally poor engine response.

It was the development of the complex breed of turbocharged 1.5 litre F1 engine in the early 1980s, with the consequent need to minimize throttle lag, as well as coming to terms with the increasingly restrictive fuel capacity limit of that period, which hastened the development of computerized injection systems.

In these systems, an electronic control unit monitors the engine's operational conditions, and also controls the injection quantity and timing in accordance with a set of parameters which identifies the ideal fuelling for every set of conditions. These functions were soon tied in with computer control of the ignition function to produce a fully integrated engine management system.

One of the first proper engine management systems to be used in F1 was developed by Zytek, the company originally founded as ERA (Electronic Racing Aids) in 1983 by former Lucas electronic engineers Bill Gibson and Brian Mason. Their first customer was Brian Hart, whose four-cylinder type 415T was being used by the Toleman team, and the ERA fuel injection control was a key factor in moderating the somewhat fierce power delivery of Hart's Harlow-built four-cylinder unit.

However, it is possible that TAG Electronics, the offshoot of the TAG McLaren group, which is a sister company of McLaren International, the World Championship-winning Grand Prix team, has taken the initiative in offering the most sophisticated engine management systems to the international motor racing world of the 1990s.

The validity of this particular theory was set to be placed on public view during the 1993 World Championship F1 season, as the McLaren Grand Prix team squared up to its opposition using Ford-Cosworth HB V8 engines kitted out with bespoke, exclusive management systems produced by TAG Electronics. In this respect the team was seeking to gain a vital edge over three rival Ford-Cosworth HB users, all of which used different management systems Benetton (Ford EEC), Lotus(Lucas) and Minardi (Magneti Marelli).

The man behind the TAGtronic engine management system is Dr Udo Zucker, the former Bosch engineer who played a leading role in developing the new Motronic MS3 management system used by both the TAG/Porsche V6 and BMW four-cylinder turbos from 1984 onwards.

This was real pioneering stuff, with no accumulated fund of knowledge into which he could tap. It was a frustrating path at times, as Zucker recounted to *Racecar Engineering* magazine in early 1992.

On one celebrated occasion the TAG/Porsche V6 refused to fire up and Zucker eventually discovered that the car's reluctance to burst into life stemmed from the fact that his Motronic control system thought that the engine had already started. In fact, what had happened was that the crankshaft had merely been nudged by an initial blip of the compressed air starter as it built up the necessary pressure to overcome the engine's compression.

The key to success with these computerized management systems is the process known as mapping. This amounts to a three-dimensional electronic plot which relates a wide range of engine load and speed to a correspondingly appropriate range of ignition and injection settings.

Fuel development

Of course, regulations concerning the composition of Formula 1 fuels have evolved progressively over the decade, but the 1991 and '92 World Championship Grand Prix seasons saw interest surrounding its composition running at fever pitch as debates over the detailed fuel specification preoccupied the pit lane.

There is a widespread but totally erroneous belief that Grand Prix cars now run on pump fuel, although it is true that the stringent regulations governing what can be used in the 1990s are a far cry from what has been allowed in the past. For example, it is a long way from the liberal rules under which the pre-war supercharged Mercedes-Benz and Auto Union machines were permitted to use volatile cocktails which blended together such additives as nitro-methane, acetone, benzole and methanol. By 1958, the sports governing body had altered the fuel regulations to prohibit alcohol-based fuels and replace them with Avgas – aviation gasoline.

The fuel companies had become understandably anxious to gain as much promotional benefit as possible from their Grand Prix involvements, but the difficulty in

The Cosworth factory from the outside, suitably complemented by the presence of a road-going Ford Escort Cosworth in the foreground.

setting and sustaining minimum standards on a world-wide basis made the switch to widely available 130 octane aviation fuel a convenient expedient.

This in turn led F1 down the route to today's F1 fuel specification which allows some freedom in terms of make-up, while at the same time laying down a limit on the octane numbers and composition, theoretically reflecting the kind of constituents available to make every-day gasoline.

Thus the rules require fuels to have a maximum rating of 102 RON (Research Octane Number), as measured by a standard industry test method (the ASTM D2699), with a maximum permissible oxygen and nitrogen content of two per cent and one per cent respectively.

No alcohols, nitro-compounds or other power-boosting additives are permitted, and the test procedure for random checks imposed by the sport's governing body FISA, following another oil industry standard (ASTM D3244), requires that a single determination of RON must be below 102.5 in order to conform with the regulations.

Within these specific parameters, the fuel design teams have to take into account safety and environmental considerations, the quantity of fuel required, its weight, plus its capacity to produce knock-free power and instantaneous throttle response.

For F1 purposes, this latter quality has to be of an enormously high order, fuel and air having to mix completely within a thousandth of a second. During the 1991 season, the McLaren team's fuel supplier Shell analyzed the time spent by one of the Honda RA101E V10 engines on full throttle, which showed that there was precious little middle ground between wide open and completely shut. For 65 per cent of the average Grand Prix distance the throttle was wide open, for 20 per cent of the time it was shut, and for only 15 per cent of the time was in transition through intermediate conditions. In this respect, while the dynamics of the engine's induction system and the efficient operation of the fuel injection are key elements, fuel volatility is of equal importance.

A full tank of fuel actually weighs in the order of 330 lb (150 kg), or about the weight of two drivers, and represents approximately one quarter of a Grand Prix car's all-up weight when it goes out to the starting grid. Obviously, for qualifying purposes the lighter the car is the better, so all competitors will compete for their grid positions with a minimal amount of fuel in their tanks, thereby permitting the drivers to exploit acceleration and braking performance to the maximum.

However, fuel consumption in the race itself depends not only on the weight of the car itself, but also on the calorific value (energy content) of the fuel, and the fact of the matter is that those fuels which offer better economy tend to be heavier. This can create something of a dilemma, posing the fuel engineers with the challenge of producing a compromise between adequate economy over a 200 mile race distance and sufficient competitive performance.

Keeping in mind the 102 RON permitted maximum, power gains by means of very high compression are clearly limited. Because engines obtain their power by oxidizing fuel, a major priority is to increase the amount of oxygen they consume. One way is by forced induction – supercharging or turbocharging – but this, of course, is now forbidden in F1.

A less obvious way is to carry the oxygen into the combustion chamber in the molecular structure of the fuel itself. Methanol contains 49.9 per cent oxygen by weight and nitro-methane 52.5 per cent. A 50/50 blend of the two can produce peak power figures some 30 per cent higher than straight gasoline, but these oxygen-based, power-boosting additives are also strictly prohibited.

Even if such fuels were permitted, rich and heavy mix-

Ilmor sump and cylinder block castings awaiting assembly at Ilmor Engineering's factory at Brixworth, near Northampton.

tures of this sort would be ruled out on the grounds of practical packaging; the load required to complete a race distance would be prohibitive from the viewpoint of the cars overall weight and the excessive fuel tank size which would be required to accommodate it.

In fact, the delicate balance between performance and economy is assessed for each individual circuit on the Championship programme well in advance, and the fuel specification most likely to meet those requirements is obviously selected.

This involves the juggling of the 100 or so hydrocarbons which go into the manufacture of everyday gasoline. Moreover, since there is no one individual hydrocarbon which is ideally suited for an F1 fuel, inevitable compromises have to be made.

Much recent controversy has centred round whether or not F1 should be seen to be taking an initiative in the switch to lead-free fuels, a move which was originally discussed with FISA by Shell, Mobil, BP, Elf and Agip late in 1990. An agreement on an unleaded requirement for 1991 was reached in principle, but eventually deferred to 1992, apparently in deference to AGIP, fuel supplier to the Ferrari team.

What particularly irked the rival fuel companies after this debate was FISA's assertion that the deferment of the lead-free requirement was done 'with the unanimous agreement of the oil companies', something it quite patently was not.

AGIP, whilst making suitably approving noises in connection with the lead-free proposal, wanted to retain the leaded specification for another year in an effort to sustain the momentum of Ferrari's engine development programme. This had come on in leaps and bounds during the second half of 1990, Alain Prost moving the famous Italian team to within hailing distance of its first Drivers' World Championship in more than a decade.

The nature of the hydrocarbon blend used by AGIP in producing the Ferrari fuel could only be guessed at by rival manufacturers, but early speculation that the team was using a power-boosting chemical additive called furan was quickly discounted. However, it certainly seemed clear that the hydrocarbon make-up of this AGIP fuel relied on sustained lead content in order to maximize its performance potential.

Some doubt was also cast over whether AGIP's leaded fuel would have been legal on other counts during 1991. Rival companies pointed to the fact that both Britain and Germany had rules forbidding the use and transport of any fuels containing more than 0.15 gram per litre of lead. In addition, FISA was also aware that from the start of 1992 all racing fuels would have to comply with EC standards requiring no lead and no more than 2.5 per cent oxygen content.

There is another key point to be made in connection with fuel development for specific engines. That required for a 1.5 litre turbocharged V6 is totally different to that for a 3.5 litre naturally aspirated V10, and again totally different to that used by a similar capacity V12. As fuel suppliers to the Honda Marlboro-McLaren team, which won the Championship four times during their five-year partnership between 1988 and '91, Shell came quickly to understand this reality as well as anybody in the business.

Shell's first experience with Honda was with the immensely successful 1.5 litre turbo V8, running with a 2.5 bar boost limit and restricted to 150 litres under the 1988 technical regulations. These McLaren-Honda MP4/4s won a record 15 out of the seasons 16 races in the hands of Ayrton Senna and Alain Prost. Roger Lindsay, Head of Fuels Development at Shell International explains:

The engine assembly area at Ilmor Engineering, a reflection of the sheer volume of production work involved in servicing both the F1 and Indycar arenas.

'Searching for more power via the fuel was not really an issue. A small adjustment to the boost could lift the power by 100 bhp. No fuel could match that. But it still had three critical jobs to do.

'The first was to keep the engine out of knock. Your road car or mine might pink as we accelerate and we may not need to worry too much about that, other than to have the engine adjusted or change to a higher octane fuel. But with the piston tops in the turbo F1 engine probably glowing with heat, a further rise in temperature that comes with knock would punch a hole like a fist through a paper bag – broken pistons, loose con-rods, loud bangs – clearly not something that the team wanted or that we were prepared to allow.

'The second was the avoidance of vapour lock, something we rarely come across in the UK these days in ordinary driving. Fuels and fuel systems have had it designed out. However, with F1 engines and fuel systems it is always a threat, with races usually run in very hot weather and with the fuel tank cell nestling close to the engine, temperatures of perhaps 50-degree C (130-degree F) or more might be expected.

'This is well above the boiling (vaporization) point of some gasoline fractions. The wrong blend will result in vapour in the fuel system causing misfire, power loss or even stoppage.

'Another aspect of vapour formation is that if the fuel tank is vented to atmosphere to stop vapour forming inside as it empties, and the fuel starts to vaporize, then it's a bit like leaving a kettle on a stove; sooner or later all the water will boil away, and unless you have been watching it closely, you won't know. That's not to say, of course, that a whole tankful of fuel would evaporate away, but perhaps enough to cause an embarrassing early run-out.

'Mileage, of course, was the all-important fuel variable, especially with the 150 litre limit. Working with a high density fuel gave the most mass for the volume available. But this time the key element in the mileage, was in the lubricant.

'Traditionally, racing engines have run on thick (viscous) lubricants – typically SAE 50 – because engineers wanted high oil pressure and protection against dilution of oil by unburnt fuel draining into the crankcase.

'My lubricants' colleagues realized that with the exquisite fuel metering and injection system, the Honda engine did not dilute the oil so badly. They combined that thought with the work then being done at Shell's Thornton laboratory to measure the thickness of the oil film between the big end bearings and the crankshaft. From that they deduced that the Honda engine could safely be lubricated with what was virtually a thin winter grade oil.

'This had the effect of releasing several extra horsepower normally consumed in oil drag and friction, or alternatively, at constant power, improving fuel consumption by two per cent. This gave the car an extra lap's worth of fuel reserve over the course of a race, and the drivers the extra confidence that went with that.

'It's worth nothing that, in round terms, about 30 per cent of the fuel energy goes into useful work (developing bhp), 40 per cent goes down the exhaust as hot gas and noise, and 30 per cent is lost as waste heat – and of that 30 per cent, 10 per cent is absorbed by the oil.'

Roger Lindsay draws comparisons with the state of the technology available 30 years ago, making the point that the lubricants used by BRM in its 1962 Championship-winning 1.5 litre V8 were required to handle about 50 kW of waste heat. In the recent range of Honda V12 engines, the oil had to absorb more than three times this heat level – 180 kW – without thickening or causing deposits.

With the end of the turbo era in 1988, Shell's team had to concentrate its efforts on the totally different fuel requirements of the new 3.5 litre V10. Lindsay continues:

'Among the characteristics we found different were the naturally aspirated V10's increased sensitivity to the volatility of the fuel. With no turbocharger to heat and mix the fuel and air, in practice we had to blend lighter and more volatile fuel. This meant a lower density fuel with poor consumption and less margin against vapour lock. In turn, this affected the octane characteristics that we could build into the fuel. The octane properties of the lighter (volatile) hydrocarbons are not as good as those of the heavier ones, so were restricted as compared to the octane characteristics provided for the turbo car.

'It was at this time that we began to consider the prospects of a fuel for qualifying. Since qualifying meant only running short distances, the limitation of fuel consumption did not apply. This meant that we had more freedom than for race fuel with respect to blending for the other aspects of performance such as power, and early in the 1989 season we produced our first successful qualifying fuel.

'Thus we progressed through another successful season, during which we had also started to take into account the effect of fuel composition on the weight of the fuel, and therefore on the weight of the car, as well as tuning the fuel for different circuits. With a full tank, the fuel weighs between 150 and 180 kg, depending on the density. That 30 kg difference is some 4 per cent of the car's all-up weight and, if it isn't needed, represents an excellent bonus for the team, especially in the early stages of a race. Fresh challenges again were presented by the Honda V12 engine in 1991 and with its arrival began the most active period of development during our years of collaboration with Honda.'

However, the regulations controlling the specifications of F1 fuels were to be thrown into confusion during the summer of 1992 when FISA insisted that the cars should use pump fuel from the Hungarian Grand Prix weekend onwards. This edict was accompanied by dark hints that detailed analysis of samples taken from certain cars at the French Grand Prix had revealed irregularities in their content.

FISA insisted on a strict interpretation of the 1992 fuel rules, but then took a closer look at its *own* rules and realized that, strictly speaking, they were not legally valid, as they had not in fact been ratified by all signatories to the Concorde Agreement. FISA then issued a clarification stating that the legal composition of fuel would be governed by regulations originally laid down in 1978.

This provoked outrage from Elf, as fuel supplies to the dominant Williams-Renault *equipe*, who felt that FISA was discriminating against them in particular, and accused the sport's governing body of disregard for the legitimate interests of the fuel suppliers and unambiguous threats to the fuel companies.

Formula 1 history is, of course, littered with examples of these high-profile disputes, but by the end of the 1992 season the biggest challenge facing Grand Prix racing was the question of formulating fresh engine regulations to take effect beyond 1995, the point at which the current rule stability is due either for replacement or reaffirmation.

Major reductions in power output must be regarded as a key element in any new formula, most enlightened designers and team owners believing that a careful programme of structured technical development should be instigated to take F1 racing into the twenty-first century with the absolute minimum of drama and controversy. Inevitably, of course, the means by which this process is achieved gives rise to even more self-generating discussion and debate.

As an example of this, Luca di Montezemolo, who took over as Ferrari President in late 1991, repeatedly found himself warning that the Italian team could quit F1 unless fundamental and wide-ranging changes are made to Grand Prix regulations to bring the sport back to its manufacturing roots. During a speech to a group of businessmen at Bologna in February 1993, he said:

> 'There is nothing forcing us to remain in Formula 1, which must change its rules absolutely to return closer to the technology of mass-produced cars. Ferrari will never stop racing, but if things do not change quickly, we could opt for other types of competition. We have reached a point where 95 per cent of the technology learned from racing has no application to road cars.'

But does it have to? Williams Technical Director Patrick Head feels that this cannot be regarded as a strict, absolute requirement. He has fearlessly stated:

> 'Formula 1 is a totally artificially generated competition, and, although I'm involved for technical reasons, ultimately the racing must be for entertainment because the people who pay for it are the sponsors and spectators.'

Charting a course which strikes the ideal balance between these two obvious requirements is certain to tax the ingenuity of F1's rule-makers over the next few years. But whatever the technology involved on a day-to-day basis within F1, the sport's enduring success in financially, politically and environmentally troubled times is almost certain to be measured against these crucial yardsticks.

CHAPTER 5

SUSPENSION DEVELOPMENT

Slightly more than 30 years ago, the transverse leaf spring had only just given way to the familiar double wishbone/coil spring configuration as the most popular suspension arrangement used by Formula 1 designers. This set-up enjoyed a long life through to the start of the 1970s, offering the benefits of a relatively low unsprung weight, which permitted a more responsive reaction to undulations on the track surface, as well as offering ease of accessibility and maintenance, and greater geometrical flexibility.

To start with it became the accepted practice to mount the co-axial spring/damper units between the bottom of the suspension upright and the top of the chassis, but eventually it became the trend to move them completely out of the airstream within the chassis, activated by extensions to the upper wishbones which acted like rocker arms. The advent of side-mounted water radiators and, later, aerodynamic side pods generally made this repositioning of the front suspension components more desirable. But that was only part of the battle.

When Colin Chapman produced the epochal Lotus 78 wing car in time for the 1977 Grand Prix season, its potential benefits were diluted due to some very untidy packaging of the rear suspension. While at the front the coil spring/dampers were tucked into niches within the monocoque walls, at the rear, in a jumble of massive uprights, inclined coil spring/dampers and exhaust pipes cluttered up the point beneath the side pods from which the airflow required a clean exit for maximum aerodynamic efficiency.

The Lotus 79 heralded a subtle change of emphasis in this respect in time for the 1978 season. The rear suspension was tucked away close to the engine/gearbox package, with only the rocker arms and lower wishbones protruding into the ground-effect venturi which now swept right through to the rear of the car. By this stage in the evolution of the Grand Prix car, suspension design was still an important element in the overall package, but the 79's aerodynamic requirements underscored the fact that the influence on other aspects of the car's performance, exercised by the way in which the suspension was packaged, had to be taken into account.

However, the adoption of inboard activated rocker arm suspension meant that the upper arm had to resist the entire download as a bending force, and as aerodynamic loadings increased with the arrival of ground-effect, the strength and cross-sectional area of these fabricated arms increased proportionally, imposing an increasing effect on the car's aerodynamic performance as well as its all-up weight. An alternative was needed.

The next step was the development of the pull-rod system, pioneered by Gordon Murray, then Brabham chief designer and now technical director of McLaren Cars. He concluded that no matter how much one strengthened suspension rocker arms, these bending movements caused them to act like undamped transverse leaf springs. His idea was to retain the double wishbone set-up, but to control wheel movement by means of a thin pull-rod running from the top of the upright to the bottom of the spring/damper.

This pull-rod system was introduced on Murray's Brabham BT44 design in 1974, after which it would take another nine years before McLaren designer John Barnard took the concept one step further, introducing a push-rod system, whereby the rod linked the bottom of the upright to the top of the spring. This system, although deemed by Murray at the time to be 'about 10 to 15 per cent heavier than a pull-rod arrangement', did have the benefit that an aperture did not have to be cut in the monocoque wall.

Rocker arm front suspension configurations survived, even on some of the fastest cars in the F1 business, through to the end of 1981, by which time it had become quite clear that this was no longer the ideal set-up for the ground-effect era.

For example, when the Williams FW07 was finally replaced after three derivatives of the original design had largely dominated the F1 scene between 1979 and '81, pull-rod front suspension, with much thinner profile wishbones, was a key feature of the replacement design.

Wind tunnel research had revealed a significant amount of aerodynamic lift being generated by the rocker arms and their associated fairings on the front suspension. The switch to a pull-rod set-up was made solely with these aerodynamic considerations in mind. Frank Dernie, then the team's aerodynamicist, recalls:

'People said that pull-rod configurations were fundamentally better for suspension than what had gone before, but in my view it wasn't any better at all. Any loaded component on a racing car should be in tension or compression, if possible, and the pull-rod arrangement merely helped us as far as aerodynamics were concerned, as well as achieving a very small percentage weight saving.'

The downforce generated by a ground-effect car varied rapidly with its ground clearance, and its effectiveness was therefore critically dependent on sustaining a constant ride height. This consideration became even more crucial from the start of 1983, when FISA decided to implement a rule requiring absolutely flat undersides from the rear of the front to the front of the rear wheels.

Accordingly, by the early 1980s, Frank Dernie concluded that suspension geometry, as such, was virtually meaningless; the suspension configuration had merely become an adjunct to the car's overall aerodynamic performance. In his estimation, once a Grand Prix car had built up a worthwhile head of speed, its handling characteristics became dominated by its aerodynamic performance and precious little else. McLaren International Technical Director Neil Oatley did not totally share that viewpoint:

'I would say that, during the 1980s, suspension geometry became nowhere near as significant as it had been, but it still had a contribution to make. But as the years go by, the geometry and layout become less and less important as suspension movement is further reduced.

'The priority for the suspension is to keep the car's aerodynamics as stable as possible, while still permitting a degree of compliance over the kerbs and bumps. Unfortunately that is, on the face of it, something of a contradiction, but it is what everybody in F1 is attempting to do. The way the actual suspension is packaged is only one element of a more complex overall package.'

Thus, by the mid-1990s, suspension design had become just one more aspect of packaging and integration of the whole design. But the requirement for an increasingly precise ride height control prompted Team Lotus to accelerate the development of their active suspension system in time for the 1987 season, when it was employed on the Lotus-Honda 99Ts driven by Ayrton Senna and Satoru Nakajima. Its purpose was to offer the driver a more comfortable ride as the ultra-stiff conventional suspension systems were producing an unwelcome by-product in terms of a terrible physical pounding for the man behind the wheel.

Developed since the early 1980s in conjunction with the Cranfield Institute of Technology – and in particular with their Flight Instrumentation Group, which had a variable feed control system under development for installation in the British Aerospace Hawk trainer – the system was based around a system of hydraulic jacks, which replaced the conventional coil spring/dampers, programmed to react and compensate in response to impulses received from a series of potentiometers and accelerometers.

Moreover, the system as used by Lotus that season had been further developed to incorporate a back-up system of secondary steel springs which would support the normal weight of the car and thus reduce the total hydraulic power requirement.

In the event of a hydraulic failure, it was anticipated that these would be sufficient to enable the driver to struggle back to the pits, rather than abandon his car out on the circuit. In fact, with a driver of Senna's consummate skill and mechanical sensitivity, it allowed the Lotus to struggle home third in the 1987 German Grand Prix at Hockenheim in just such a stricken condition.

Five variable-flow valves provided an interface between the electronic control systems and the hydraulic activators. One of these valves was fitted to an outlet of the pressure pump driven via a mechanical coupling from the tail of the left-hand exhaust camshaft on the Honda V6 engine, the other four being mounted on each suspension unit where they converted the electronic impulses into hydraulic movement.

It was extremely difficult to assess the benefits of the Lotus 99T's active suspension within the wider context of how the car fared against its opposition. It arrived on the scene when the F1 turbo engine regulations imposed a 195 litre fuel capacity maximum, and due to the Lotus's poor aerodynamics, Senna was unable to get on terms with the Williams FW11Bs which shared the same Honda V6 engine, and were driven by Nelson Piquet and Nigel

Mansell. The system also added over 22 lb (10 kg) to the cars all-up weight, and there was inevitably a slight power loss resulting from driving the hydraulic pump.

At the same time, Williams was busy developing an active-ride system based round a concept pioneered for Automotive Products by their engineer Bob Pitcher. It was a less complex arrangement than that on the Lotus, relying on fewer sensors on the car, but it did help Nelson Piquet to victory in the 1987 Italian Grand Prix at Monza.

These Williams active-ride cars employed an accumulator which was charged by a hydraulic pressure pump driven, like the Lotus, from the left-hand exhaust camshaft of the Honda engine. Normal suspension movements were handled by struts and gas springs mounted in series, each one also interacting with an electronically commanded valve which permitted pressured hydraulic fluid to move in and out of the strut in order to control its length.

A rheostat mounted in parallel with each strut then signalled changes in length to a central computer which, acting upon this input, was then programmed to respond by sending messages to the hydraulic valve blocks designed to keep the car level, cancelling roll and pitch, and to maintain a constant ride height.

This system was then incorporated into the basic design of the 1988 Williams-Judd FW12, but was troubled with such a hideous level of unreliability – due to engine vibration problems and localized over-heating – that the car was revamped to incorporate a more conventional springing medium by mid-season. As we shall see later, however, this was far from being the end of the active suspension story.

Another element which has exerted an influence on suspension packaging over recent years has been the premium placed on the smallest possible frontal area for the F1 chassis. This has prompted designers to consider several options as far as positioning the front spring/dampers are concerned.

John Barnard, on his first time round as Ferrari's Technical Director, opted to mount the front shock absorbers on the type 640 design above the driver's feet, longitudinally down the front of the monocoque. Before long, those designers opting for conventional spring/dampers within the monocoque were left in a small minority. Even McLaren had abandoned this configuration by 1990, the spring/dampers on the 1991 MP4/6 being mounted ahead of the cockpit atop the monocoque. In 1989, Tyrrell

The rear end of the 1988 World Championship-winning McLaren MP4/4 reveals a tightly packaged layout similar to the March, although the coil spring/dampers are still mounted vertically, operated by rockers atop the push-rods which can be seen just to the rear of the oil tank, level with the leading edge of the rear wheels.

designer Harvey Postlethwaite followed a similar path, but instead opted for a single spring/damper linked directly to the push-rods on either side of the chassis.

With the large front tyres employed on contemporary F1 cars being required to generate an enormous amount of grip, the necessity to work them very hard in order to generate their working temperatures had led to many teams running very stiff front anti-roll bars at most circuits.

By running what amounts to a solid anti-roll bar, the effect produced what was virtually a 'monoshock' arrangement by constraining the conventional twin spring/damper units to move together. It thus made sense to replace this arrangement by a single spring/damper, connected to a single rocker attaching to both front push-rods, with a resultant very significant weight saving.

There were some snags, however, in operating such a solid axle concept with loss of front-end grip on some circuits, and particular problems over a bumpy track surface where both front wheels would be lifted momentarily if just one of them hit a bump.

Andy Brown, in 1991 responsible for engineering Martin Brundle's Brabham-Yamaha BT60, lucidly recalled his own first experiences of such a system in a technical

Rather than mounting the coil spring/damper units vertically within the walls of the monocoque, it became increasingly F1 practice to position them longitudinally atop the monocoque immediately in front of the driver's cockpit during the early 1990s, as this shot of the 1992 Ligier-Renault JS37 amply illustrates. This facilitated better access as well as enabling the cross-section of the monocoque to be reduced in size around the footwell.

article published during the following season:

'An advantage of this (monoshock) system that we discovered was that if you were to try riding a kerb with a standard suspension system, the effect of the inner wheel hitting the kerb would cause the outer wheel to be lifted also, due to the bump effect of the kerb being transmitted to the outer wheel via the anti-roll bar.

'With the monoshock system, when the inside wheel rode the kerb, the effect of the rocker sliding laterally across the car would be to push the offside wheel down, not lift it as before, and hence improve the car's grip under these conditions.

'It also took us a while to sort out the system's characteristics, but from the mid-point of the season we were always challenging for Championship points, as long as the track was smooth. The problem for us at bumpy circuits was that the factor of velocity ratio (the relationship between wheel and damper travel) had not been considered when switching from two dampers to the monoshock system. On the Brabham, we therefore effectively had to double the spring rate fitted to the front as, quite simply, one spring was doing the job of two. That led us to running as much as a 5,000 lb/ft spring at some circuits.

'Whilst this may not seem too bad, for as far as each wheel is concerned it is controlled by a 2,500 lb spring as before, the problem was in the shock absorber settings. The dampers need a fair amount of shaft velocity if they are to do their work, plus a significant difference in shaft velocity over the bumps than for the rolling and pitching movements of the car due to driver input, if the dampers are to be used to the full in controlling both types of motion.

'If the velocity ratio is not changed, then the damper settings have to be increased. As the low speed forces (the pitch and roll motion of the chassis) are usually controlled by varying the number of bleed holes in the damper piston, and as the number of holes is usually very small on a dual shock absorber system, it may not be able to increase the low speed forces sufficiently, whilst still allowing enough fluid to bleed past the damper, without having to lift the piston shim pack.

'The shim pack is a series of pre-loaded washers which are used mostly to control high speed forces on bumpy circuits, and hence you start to lose control of these forces if the shim pack starts to be used primarily for low speed control, making the suspension feel very hard over the bumps. Therefore careful consideration needs to be given to the velocity ratio when defining the figure required for a monoshock system.'

By the start of the 1992 season many F1 teams were using, or experimenting with, monoshock suspension systems, but then the Williams team suddenly redrew the parameters of F1 car performance by introducing the latest version of their active suspension system on the new Williams FW14B. Ever since abandoning the system on the Judd-engined FW12 midway through 1988, the further refinement and eventual reintroduction of the concept was always on the team's technical agenda, albeit perhaps on the back burner.

Throughout 1990 and '91, the team's test driver Damon Hill put the fledgeling system through an intensive development programme, running many thousands of miles away from the Grand Prix spotlight. It was not until the early weeks of 1992 that Patrick Head, having discussed the matter at length with Frank Williams, Nigel Mansell and Riccardo Patrese, finally signalled a total commitment to using the system in a full World Championship programme. It was a factor destined to give Mansell a vital technical edge during a season which saw him romp, scarcely challenged, to the World Championship.

Whilst relying on conventional springs and dampers to soak up the bumps, the Williams active system was primarily a means to stabilize the under-car aerodynamic performance by means of computer-controlled hydraulic rams operating as extensions of the suspension pushrods.

Before the start of the 1992 season, Patrick Head was understandably reluctant to spill the beans in any great detail about his revised suspension, privately believing it to offer 'a significant performance advantage'. But he didn't want to make any wild predictions in public about its potential. Obviously he still felt very aware of the failures of 1988.

On reflection, Head admitted that his design department bit off more than it could chew with the active ride system in 1988. As he explained:

'The system had worked well on the Honda turbo car in 1987, and we believed it had to be the way to go with the Judd-engined car the following year. However, the difference was that the Judd car was significantly smaller, which made the packaging of the system – keeping things away from hot places and so on – much more difficult. Also, there was a great deal more vibration from the engine which certainly didn't help.'

He continues to make the point that, in terms of responsiveness, the Williams FW12 seldom did the same thing twice. That was a significant negative factor, because Mansell, who was the team's number one driver that year, was an instinct driver and instinct requires regularity of feedback. Head recalls:

'Pretty understandably, I think, Nigel wasn't very enamoured of it. He always had a basic mistrust of active systems anyway, and that went back to being dumped in a sea of hydraulic fluid at 180 mph, something he had a lot of during his early years at Lotus...'

Yet even Mansell was convinced by the start of 1992, and Williams was confident that the basic development problems were behind them. Head:

'The system we were to use in 1992 was based on what we had in 1988, but the software approach is completely different. This time we had a lot of double-sensing, lock-out systems and so on. If, say, the pump fails, a system of solenoids clunk together and lock it, so the suspension cannot collapse.'

Nevertheless, there was the unspoken prospect of one of the struts fully extending at the wrong moment, an untimely occurrence which pitched then-Williams driver Thierry Boutsen into a huge accident whilst evaluating the system at Estoril during the closing months of 1990. Again, on this front, Head remained confident that the prospect of such a repeat performance had been substantially reduced.

The use of active suspension systems in F1 has now become a key factor in controlling the behaviour of undercar aerodynamics, and going into the 1993 season any team which aspired to front-running status had such a system ready to race – or at least well advanced in the development stage. However, by then an increasing body of informed thought had concluded that the business of 'bottom-dragging' F1 cars was a complete nonsense, and that steps would have to be taken to initiate major rule changes very quickly.

CHAPTER 6

FORMULA 1 TRANSMISSIONS

In common with every other aspect of racing car technology, the past couple of decades have witnessed enormous strides in transmission development, both from the viewpoint of the technology involved in component manufacture and from the standpoint of integrating the transmission package into the overall design. Go back 20 years and just as you will find that most of the cars were powered by Cosworth DFV engines, so most of them transmitted their power to the rear wheels by means of gearboxes manufactured by Hewland Engineering, the specialist Maidenhead-based company which still thrives as a transmission supplier to race car manufacturers in a host of different categories.

In many ways, up to the mid-1980s, the story of F1 gearbox development – away from Ferrari – was the story of Hewland Engineering. Founded in 1957 by Mike Hewland and one assistant, in a shed opposite Maidenhead railway station, the company was established to produce a wide variety of sub-contracted components for various racing teams. Their move into transmission work began two years later when the UDT/Laystall F1 team, which was run by Stirling Moss's father Alfred and his business manager Ken Gregory, approached Hewland with a request for him to design and manufacture a transaxle system for the 2.5 litre-engined Coopers.

This was followed in 1961 by a commission from Lola boss Eric Broadley to produce a race-ready transmission based on a proprietary gearbox, the Hewland Mk 1 transmission being based around a Volkswagen gearbox for these purposes. Hewland would eventually arrive full-time on the F1 scene with its HD gearbox fitted to the 1966

World Championship-winning Brabham-Repcos, although this was not really man enough for the job of handling 300-plus bhp, and Brabham subsequently commissioned Hewland to produce a heavier gearbox, dubbed the DG (for 'different gearbox'), which would eventually do the trick.

However, while the Cosworth DFV's power spread only ranged from 370 to 470 bhp across a 15-year span – making transmission requirements relatively quantifiable from year to year – the advent of the 1.5 litre turbo era at the start of the 1980s significantly altered this particular status quo. Whilst individual Grand Prix teams would now concentrate on developing their own subtle modifications to what were essentially off-the-shelf gearboxes, the massive leapfrog in power produced by the turbo era also promoted many to manufacture their own special outer casings. Nevertheless, even as late as the 1986 season, every British F1 team with the exception of Brabham was using Hewland gears inside its transmission.

The Hewland FGB gearbox, which formed the basis of many F1 transmission systems through the 1980s, owed its development to the original FGA gearbox which had been introduced as long ago as 1969.

One of the first departures from the Hewland route amongst British teams came in 1981 when the requirements of ground-effect aerodynamics led Gordon Murray to explore the possibility of using a transverse gearbox/final drive unit for the Cosworth DFV-engined Brabham BT49, which had hitherto been relying on an Alfa Romeo transmission casing equipped with Hewland ratios.

For that year's United States Grand Prix West at Long Beach, one of the team's cars was fitted with a transmission produced by American specialist Pete Weismann, previously better known for his transmission work on Indianapolis and Can-Am cars. The new Brabham gearbox could be fitted with either five or six speeds, the gear shafts running transversely across the car to enhance under-car aerodynamic potential by keeping the rear end of the chassis as smooth and uncluttered as possible.

The directional change from the crank axis to the gear shafts was effected by low friction, straight-cut bevel gears, and on the right-hand of the casing was a detachable plate through which the gear cluster could be withdrawn in order to change ratios. This transmission was packaged in conjunction with inboard-mounted coil spring/damper units, activated by rocker arms, which were mounted vertically behind the ultra-slim transmission. The oil tank, feeding a lubrication system which was common to both engine and gearbox, was mounted within the bell-housing. Murray recalls:

'The whole package was extremely light, rigid and compact, but we really needed more facilities and personnel to develop it to its optimum pitch. Ideally we would have required a separate factory, with about a dozen technicians on the project full-time to make it work properly. By that stage, F1 gearbox development had become a very time-consuming and expensive business.'

Prophetic words indeed from the Brabham chief designer...

A derivative of this system was originally envisaged by Murray within the design of the ground-effect Brabham BT51 which was originally envisaged for the 1983 season. However, this had to be scrapped after the implementation of the new rules requiring flat-bottomed cars.

That said, through the period 1983–5, when turbo engine development was being carried out unfettered by restrictions on turbo boost pressure, the Brabham transmission benefited from input from Weismann, who had been retained as a consultant by the British team. As Murray points out, with upwards of 1,000 bhp being developed by the BMW turbo in qualifying trim, the transmissions would prove extremely marginal on several occasions, but they held together long enough to enable Nelson Piquet to win the first turbo World Championship in 1983.

Weismann was again called upon by Murray in 1986 to produce a new gearbox for the striking low-line Brabham BT55, the car with which the South African-born designer hoped to reinvigorate his reputation as an innovative and trend-setting arbiter of F1 design. To cater for the BMW four-cylinder engine installed at an angle of 72-degrees, Weismann produced a bevel drive transmission incorporating an 'overdrive' seventh forward gear. This, it was hoped, would enable the BT55 to be geared to produce startling acceleration as well as enviable top speed. Unfortunately the project was sunk due to a variety of unrelated technical problems, so an assessment of the transmission's worth was difficult to arrive at.

Hewland internals from the heavier-duty Hewland DGB gearbox were being employed in the transmissions of most British-based turbo cars, but Mike Hewland was extremely concerned about the torque loadings rather than the outright increase in power. He would reflect:

'The massive increase in torque we were having to deal with jumped from about 250 lb/ft in the heyday of the Cosworth DFV to as much as 400 lb/ft in the case of the most powerful turbos. When I think of our tiny first gear ratios, which could be covered by the palm of your hand, and then see all the ferocious power which goes through them at the start of a Grand Prix, I sometimes wonder how anybody gets off the grid at all.'

Tyrrell would be another to take the decision to make its own gearbox casing as long ago as 1979, primarily to permit the inboard packaging of the coil spring/damper units in the interests of optimizing the type 009's ground-effect aerodynamics. Tyrrell could hardly have imagined the irony involved in this development, nor the long service it would eventually see in the hands of a rival team!

McLaren boss Teddy Mayer did a deal to use this casing on the McLaren M30 in 1980, and it was then used on the succession of highly successful McLaren MP4 derivatives under John Barnard's technical stewardship.

'With all the problems involved in making the new carbon fibre chassis, there just wasn't the time or the resources available to tackle a new casing', he explained. As a direct consequence, a deal with Tyrrell was concluded to make further modifications to the casing, and this would last the TAG turbo-engined McLarens right through to the end of their competitive life in 1987. By 1984, however, McLaren had switched away from Hewland internals and was having its own gears manufactured on a bespoke basis by either Emco or Arrow gears in the USA.

The question of transmission packaging would become an increasingly central issue in the business of Grand Prix car design through the 1980s, as designers sought to retain optimum wheelbases while at the same time accommodating a variety of fuel loads, chassis balance requirements and footwell regulations.

In 1988 McLaren would switch to Honda 1.5 litre turbo engines, and while retaining their conventional transmission layouts, with the gear pack sticking out behind the crownwheel and pinion, had to make other changes to accommodate the Japanese engine. Due to the Honda RA168-E's low crankshaft position, it was necessary to introduce a new three-shaft gearbox to 'step up' the drive to a level which would minimize potentially troublesome drive-shaft angularity.

This was developed by two of Gordon Murray's old confederates, Weismann and David North, and while the configuration revealed a slight power loss, it proved extremely reliable through the 1988 season, in which McLaren would win a record 15 out of 16 Grands Prix.

As the decade progressed, it was somewhat ironic that Hewland would be eclipsed as the leading player on the F1 gear stage by the Finchampstead-based Xtrac organization. The founder of this company, Mike Endean, had himself worked for Hewland for 16 years between 1968 and '74. The story of his company's rise from a one-man band, having its roots in a single freelance commission, to the point where it would supply transmission systems to virtually every Grand Prix team in the business, would become one of the most outstanding F1 technical success stories of the decade.

Back in the late 1960s, Endean himself raced, first with a home-brewed clubman's sports car, and subsequently in Formula Ford with a Dulon he shared with Alan Baillie. 'It was his car, but I supplied the gearbox, one of the first Hewland Mk 8 transmission', he remembers.

In his professional life, Endean eventually found himself outgrowing any challenge he could find working at Hewland. The Maidenhead company's FGB unit had basically been the last off-the-shelf gearbox to win a World Championship, when Alan Jones took the Williams FW07 to the title in 1980. From then on, the demand which increasingly built up for bespoke gearboxes opened up a whole new market for enterprising specialist concerns.

However, Endean makes the point that it wasn't a question of Hewland lacking the initiative to tackle such tasks, more a lack of expectation on the part of competing teams that a gear manufacturer would become involved in making such refinements as special casings and dry-sump systems.

It was a personal request for a rallycross transmission system which started Endean down the road to independence. After building a two-wheel drive gearbox for his Ford Escort rallycross car, Martin Schanche persuaded Mike to build the 4WD system he so urgently needed to tackle the Audi Quattro challenge in this specialized branch of the sport.

Endean started carrying out this work in his spare time, but it wasn't long before he was facing so many requests for the XTrac 4WD system that he decided to go it alone, moving into a small workshop near Wokingham station, with a staff of only three assistants, at Christmas 1984.

XTrac gradually moved into F1 during the mid-1980s when, as Mike Endean recalls, 'We started making any parts that didn't actually drive the car along, with the exception of shafts. That is to say things like clutch shafts, drive-shafts, oil pumps, reverse gears and so on.'

In 1988 XTrac drew its first F1 transmission casing, for the now defunct Onyx team, since when investment continued at a high level, motivated by Endean's belief that such facilities as its own heat treatment plant was essential if XTrac was to start the complete manufacturing of its own gears and capitalizing on the obvious market which was opening up in F1.

Prior to the installation of the heat treatment plant, XTrac had supplied some gear sets to McLaren for testing purposes in its first, transverse-mounted, six-speed box. For packaging reasons, of course, a transverse gearbox lying ahead of the rear axle line is theoretically better from the viewpoint of weight distribution and, of course, leaves room for the crucial rear aerodynamic diffuser panels, but Endean admits that he is still undecided about the pros and cons of transverse/longitudinal boxes from the pure efficiency viewpoint. During the summer of 1992, he explained:

'We are actually involved in a detailed programme on our test rigs. Because more of the outside surface of a transverse box is designed with a view to aerodynamics, I think that the longitudinal box is probably slightly more efficient, although I can't prove that. It's just the thought that more oil has got to be in contact with various things in a transverse box than in a longitudinal, bearing in mind that you are using a dry sump system in both.'

By this stage in the company's development, establishing a series of static rig tests had come to be regarded as essential by XTrac, so that in the absence of its own test car, the company could start to establish its own data base.

By 1989, rallying work still represented about 85 per cent of XTrac's business, but the F1 side was expanding steadily. The company was even asked to manufacture some prototype components for the radical new Ferrari 640 design which was being conceived by John Barnard at Maranello's British design studio, GTO, near Guildford.

Endean and his general manager Peter Digby remember

that one of the front-line F1 teams was understandably scrupulous about the way in which it went about assessing the quality of XTrac's components. At the start of 1989, when the team commissioned them to make its first batch of gears, it sourced some alternative supplies from a US manufacturer. At the end of that year it did the same, just to check that XTrac's standards hadn't shifted.

By the start of 1991 it was clear that XTrac's F1 business was increasingly dramatically, and after researching the matter for around 12 months, the company made what it regarded as a crucial £1 million investment in a computer-controlled Klingelnberg bevel cutter, in order to manufacture its own crownwheel and pinions in hours. This was done even in the full knowledge that, once the order was placed, it would take another seven or eight months before it was delivered.

Thus, by the end of 1991, XTrac had to supply Finnish-made crownwheel and pinion units to six or seven of its F1 customers. Understandably concerned about quality control, XTrac was airlifting the material out to Finland where the soft cutting was done, then flying it back to Finchampstead for heat treatment, back to Finland again for heat treating, and then back to England for final delivery.

Even when the new machine was delivered and F1 customers were literally banging on the door wanting crownwheel and pinions, XTrac squared up to the problem with complete frankness. Peter Digby remembers:

'We just decided that we couldn't have customers sitting on the grid in March 1992 saying that XTrac had let them down on crownwheel and pinions, so we said to many of them, "look, this is a new machine and it might have some problems, so we can't allow you to come to us yet because we need to prove it".'

In the event, the XTrac machine cut its first set of crownwheel and pinions in November 1991 and has proved very successful ever since.

In 1988, however, revised footbox regulations which required the positioning of the driver's feet to be completely behind the front axle line prompted a fresh challenge from the overall chassis packaging viewpoint, if the designers were to retain an optimum wheelbase. In addition to 'wrapround' fuel cells (which would cause potential problems with the Ferrari 640, as explained elsewhere) designers now began experimenting with either transverse-mounted gearboxes or inboard gear packs during the 1988 season.

The former method would be adopted by Williams for its Judd V8-engined FW12 during the interim 1988 season, during which teams could make the choice between 3.5 litre naturally aspirated engines, as Williams did, or soldier on with 1.5 litre turbos, using a reduced 2.5 bar boost pressure and a 150 litre fuel capacity restriction.

Williams transmission engineer Enrique Scalabroni worked closely with Technical Director Patrick Head on the development of a neat transverse gearbox, positioned ahead of the rear axle line, and utilizing what were basically Hewland DG gears which had been specifically manufactured to Williams's own specification, with narrow gear teeth and thinner discs.

Benetton, meanwhile, packaged their new 3.5 litre Cosworth DFR-engined B188 with a gear pack positioned longitudinally ahead of the rear axle line. This was also a totally new transmission which made use of smaller gears thanks to the lower level of torque developed by the DFR engine as compared with the 120-degree 1.5 litre turbo V6 which had been employed on the Benetton B187. This configuration would be followed by the rival Leyton House March team the following year, basically enabling them to adapt many of the components used on the previous year's conventional transmission. Unfortunately the new arrangement suffered from an insufficiently rigid transmission casing, which failed under load to produce a spate of gearbox problems early in the season and led to slight thickening up of the walls as a result.

In retrospect, several F1 engineers have concluded that these systems represented the worst of all worlds, the casing requiring at least one more split line between the engine and final drive in order to facilitate access for ratio changing.

In practical terms, this access problem becomes enormously difficult, mechanics being obliged to remove the chassis underbody and most of the rear suspension before the main case can be split and the gears extracted.

Moreover, if one considers the slimness of the in-line configuration and adds that to the greater distance required between the rear face of the engine and the final drive, the overall torsional rigidity of the chassis must be suspect.

This configuration had first been tried by Robin Herd on the abortive March 721X back in 1972, the designer attempting to produce a car with the lowest possible polar moment of inertia. At the time March was involved in a complicated deal with Alfa Romeo for the use of the Italian team's singularly uncompetitive V8 F1 engine, and the gearbox used by Herd for the 721X was inherited from the Alfa T33/3V8 sports car. Its main problem was that it had a hopelessly slow change – and the rest of the car wasn't up to much either – so there was no way a hard and fast conclusion could be reached as to whether or not its gearbox configuration could be regarded as a plus point.

The trend towards transverse gearboxes continued the following year, with McLaren making the switch mid-season on its new Honda V10-engined MP4/5 challenger. From the viewpoint of aerodynamic packaging, this gave the engineers more scope for the development of the crucially important diffuser panels, but the down side was the difficulty involved in producing a slick, crisp gearchange for the drivers, due to the problems involved turning the gear selection motion through 90-degrees.

This was initially achieved by means of bevel gears and levers, but in an attempt to make this more precise, a combination of slider and rockers have been introduced, doing away with the need for gears, and thus minimizing the problems involved in the system.

However, the most ambitious step forward in terms of F1 transmission would be instigated by John Barnard, and focus on methods of activation rather than basic configuration.

Equipped with a conventional clutch for use only at the start of a race, Barnard's Ferrari 640 had a semi-automatic shift mechanism controlling its seven-speed longitudinal box. There was no conventional gearchange, merely butterfly levers behind the steering wheel. The right-hand stalk was for changing up, the left for changing-down, and the driver's fingertip pressure sent an electronic impulse via a control box to an otherwise conventional, but hydraulically operated, gearbox.

Initially, the one potential operational drawback was that the driver could only change down sequentially, without missing out gears. The system's biggest advantage was that the clutch did not remain disengaged for a millisecond longer than necessary, thereby facilitating changes which could be completed significantly faster than could be achieved by a driver with a conventional system. There were also secondary benefits, such as a reduction in manufacturing complexity with the elimination of the normal gear linkage, and a slight aerodynamic gain from the resultant narrower cockpit.

Instead of the selector rods being moved in and out by means of a mechanical linkage connected to the gear lever, an electronic solenoid (or hydraulic actuator) is attached to each rod to initiate the necessary movement. With the number of actuators required, such a system runs the risk of imposing something of a weight handicap, and also calls for a complex electronic brain.

Transmission-related problems beset the Ferrari 640 after its victory in Brazil, but it quickly became apparent that these were not fundamental shortcomings. Hydraulic pump failures, broken wires to the electro-valves within the control system and other minor irritants drove the team to distraction for the first half of the season. The car was also bugged by an unrelated vibration problem which took its toll on the V12's alternator and needed several races to conquer, but from the French Grand Prix Mansell finished in the top three in five consecutive races.

For 1990, the Ferrari 642 remained basically unchanged in concept, although Barnard relinquished his position with Ferrari to start up a new research and development department with the Benetton team. Further developments to the semi-automatic shift involved a revised layout for the electro-hydraulic control valves and the provision to programme the system for non-sequential down-changing, a facility perhaps appreciated more by new team recruit Alain Prost than sitting tenant Nigel Mansell. The Englishman preferred the proven security of changing down, gear by gear, after testing the revised system early in the year. The programming could be changed in a few minutes, by external adjustments to the softwear programme, and a dual system which offered the driver the option of either configuration was tried later in the year at Estoril.

When Williams developed its electro-hydraulic gear-change mechanism for the start of the 1991 system, Patrick Head opted for a motor cycle type, sequential gearshift mechanism, basically involving a rotating drum which, when rotated in one direction, went up through the box, and down through the ratios when rotated in the other direction. This involved a much more straightforward control system than that initially employed on the Ferrari.

In 1992, McLaren went a step further with the introduction of its new semi-automatic transmission, based on an electronically controlled hydraulic gear selection mechanism, developed by McLaren in conjunction with its associated company TAG Electronics.

Devised by Honda, the throttle mechanism incorporated stepper motor actuation, a fly-by-wire ECU (electronic control unit) which monitored the position of the throttle pedal and instructed the motor accordingly. Not only can throttle actuation be faster and more precise by dispensing with the mechanical linkage between the pedal and the engine, but the throttle ECU can also take into account the requirements of the semi-automatic system.

The biggest advantage of all these systems is, of course, the most obvious. In removing the need for the driver to take his hands off the steering wheel, he can retain more control over the car for a greater portion of each racing lap. Yet, paradoxically, there are those who believe that this acceleration in technology is in danger of minimizing the role of the driver.

After a season handling the high-tech McLaren-Honda MP4/7A – with its fly-by-wire throttle actuation, semi-automatic transmission and traction control systems – triple World Champion Ayrton Senna took the opportunity to sample a Penske PC20 Indy car whilst assessing his career possibilities for 1993. He was pleasantly impressed with what he discovered:

> 'It was a big change for me which I enjoyed very much because I like the sort of challenge where you have to get used to completely new things in a very short space of time. It was tricky to drive. It was nice to get back to a manual gearchange mechanism; you get spoiled by pushing buttons and getting used to driving without lifting off the throttle. And remember, we don't have clutches in F1 cars any longer [referring to the McLaren]; the engine systems are all synchronized with the transmission. They change up and down without you even thinking which gear you want. So it was a big difference for me to suddenly go back to a car that you have to work in, to do the things you're supposed to do.'

On the Indy car racing front, Penske completed 18 months' development on an electro-hydraulic, semi-automatic shift, but just as they were ready to take it racing, it was banned. 'A shame for us', remarked team driver Emerson Fittipaldi, 'but the decision was good for racing.'

Ditto traction control systems. 'We tested a lot with traction control', explained 1992 Indy car champion Bobby Rahal, 'but then they banned the things at the end of the year – and I'm glad they did. Stuff like that takes too much away from the driver, which makes it less of a sport, in my opinion.'

Therein lies the enduring dilemma. Has the complexity and technical refinement of contemporary transmission systems in Formula 1 significantly reduced the value of driver input? And, if so, is this a retrograde step? Most observers believe that Formula 1 stands at a crossroads in the 1990s. Will it allow technical development to continue unfettered to the point that a computer could almost replace the driver? Or will it follow the Indy car route of catering primarily to a huge television audience, tailoring its technical requirements to the demands of close racing, and nothing else? Compelling arguments can be advanced to back up either viewpoint.

CHAPTER 7

DATA LOGGING

On one memorable occasion back in 1954, during the Rouen Grand Prix, the late Mike Hawthorn was so worried that he had over-revved one of his works F1 Ferraris that he fumbled around behind the instrument panel in an effort to re-set the 'telltale' needle on the rev counter.

It was a wasted effort! As things transpired, the team manager Nello Ugolini was so furious about the lower rev limit which Mike had been using as a matter of course that he received a severe bawling out. Hawthorn later wondered why he had bothered. If he was going to be told off, he mused, he might as well have made it worthwhile by leaving the rev counter telltale at its original, astronomic level.

A decade or so later, the dynamic Austrian star Jochen Rindt gained quite a reputation for extracting the utmost from the machinery, in particular his F2 cars. One of his mechanics recalls poking his head into Rindt's cockpit after a somewhat frantic practice run and inquiring what sort of reading he had been getting from the oil temperature gauge. He received a reply which seemed satisfactory. Until, that is, he saw Jochen grin and mutter 'That was a good guess, wasn't it?' He had been far too wrapped up in the business of driving to pay much meaningful attention to the gauges in the cockpit.

Today, the whole driver/pit crew relationships is dramatically different. In days gone by, drivers could keep quiet about what was going on in the cockpit if they felt it was prudent to do so. But in the 1989 Italian Grand Prix at Monza, to take an example, the Honda engineers staring at a row of computer read-outs in the pit garage appreciated that Ayrton Senna's McLaren-Honda, leading the race, was facing dire mechanical trouble even before any warning began to indicate on the car's cockpit instrumentation.

The sophisticated data logging and telemetry systems which were monitoring every aspect of the Honda V10's performance throughout the race had indicated there was nothing to be salvaged. The McLaren's engine eventually expired in a spectacular cloud of oil smoke, and as he pirouetted gently to a halt on his own lubricant, Senna could reflect on the reality there had been nothing in terms of cockpit control adjustment that he could have made to retrieve the situation.

Of course, that is not always the case. In the 1991 Mexican Grand Prix, Nigel Mansell's Williams-Renault FW14 initially pulled away into the lead before experiencing an unexpected drop-off in engine performance. Suddenly his team-mate Riccardo Patrese was pulling on to his tail, before overtaking and disappearing into the distance. In the days before telemetry systems were widely used, Mansell would have been obliged to live with his problem and put up with it until the end of the race.

On this occasion, detailed monitoring of the engine performance computers in the pits gave a clue to the possible cause of his power loss. Although it seemed of marginal significance at the time, Renault Chief Engineer Bernard Dudot did notice that Mansell's engine was running fractionally hotter than Patrese's. After a brief discussion amongst the engineers on the pit wall, Nigel was told on the radio link to make some minor adjustments to the cockpit fuel mixture control. Shortly afterwards, Mansell reported that the engine felt better and he picked up his lost speed, finishing barely a second behind Patrese at the chequered flag.

In reality, last minute detail adjustments to the car's chassis set-up on the starting grid could have played an equally significant part in Mansell's late race recovery, as indeed could the fact that easing his pace in the middle of the race helped conserve his tyre wear. Yet the central point is that the technical operation of a Grand Prix car today relies on the interaction of many specialist elements, and the contribution made by telemetry and data logging systems certainly played a crucial role for Mansell in this particular event.

A fortnight earlier, the telemetry had assisted in getting to the bottom of the last lap retirement which cost Mansell victory in the Canadian Grand Prix at Montreal. Although the details recorded on the RAM card – the quick access method by which information from the onboard VCM (Vehicle Control Monitor) can be removed from the car for further analysis – had mysteriously been wiped by the time the FW14 was retrieved and returned to the pits after the race, Dudot had seen enough on the fuel consumption graph to know that Mansell had not been pressing consistently hard in the closing stages of the race, and could thus immediately rule out the possibility that the car had run out of fuel.

In fact, the problem had arisen when Mansell allowed the revs to drop, allowing the engine to stall at the precise moment the electro-hydraulic gearchange had balked slightly and found itself in neutral.

The need for telemetry and data logging systems in F1 really gained momentum with the advent of the temperamental and complex 1.5 litre turbocharged engines in the early 1980s. In early 1981, during the development of the four-cylinder M12/13 single turbo F1 engine, BMW engineers monitored charge air and engine temperatures, turbocharger boost pressure, and the functioning of the electronic ignition and fuel injection systems by means of radio signals from the car to computer read-outs in a small van situated behind the pits.

Since then, the systems have become much more sophisticated, but the problems of installing and operating such ostensibly delicate equipment remains one of the overriding challenges to any data logging system. In addition, the need to access the information very quickly indeed has been the focal point of much attention over recent years.

When Team Lotus was operating an active suspension system on its Honda-engined cars during the 1987 season, one of the problems encountered was that the sheer volume of information available from the data logging system threatened to overwhelm the team's capacity to analyse it. They could only really scratch the surface of the information at their disposal when it came to probing it in depth.

While the active Lotus 99T was standing in the pits, an engineer's keyboard could be plugged in to input adjustments to the onboard computer programme, but the extracted data could also be fed into a larger capacity computer in the pit garage, from which it was possible to select whatever graphic or numeric display was required to view the car's performance in minute detail at any point round the circuit.

By the use of the memory (RAM) cards, information can

today be removed from the onboard VCM very quickly indeed, without having to delay the car in the pit lane for longer periods while the information is dumped into the larger computer.

As an example, a 1 megabyte card, transferred at the maximum rate a PC serial link can operate (115,200 baud) will take just over 1 minute 40 seconds to upload, a luxury which may clearly be unacceptable during a quick pitstop during the frantic intensity of an hour-long official F1 qualifying session. On the other hand, a memory card means that access time to the car is literally the few seconds needed to remove the full card and replace it with a fresh one.

As a typical example, such a data logging system was used by the Canon-Williams team to monitor the chassis and semi-automatic gearchange performance on its FW14 chassis.

In the case of the RAM card system, the frequency of recording determines how quickly the memory capacity is used up. 'In practice, we may use up to 50 Hz, which is to say 50 pieces of information a second', explains Patrick Head, 'in which case a RAM card will have used up all its capacity in about two laps.'

Inevitably, the intensity of data gathering has a higher priority in practice and qualifying than it does in a race. The reason for this is simple. During practice, the engineers are attempting to identify the optimum settings, both in terms of engine and chassis adjustments, which will serve to produce the most competitive possible performance when it comes to the race. Once the race is underway, details of vibration levels, for example, are of less immediate significance.

There are also facilities on systems like those used by Williams for the drivers to switch off their data logging equipment by means of a cockpit control if something happens which they think the engineers might want to examine in detail. By turning off the system, the recorded details are frozen and the continuous recording function – whereby the RAM cards wipes out earlier information and replaces it with the most recent data – is immediately suspended.

During the race, the data logging capability can be reprogrammed so that the RAM card records every detail of a car's gearchange performance throughout the race. In this configuration the system automatically switches itself on and off a split second before and after every gearchange, for which function the card's memory capacity is sufficient for the entire race distance.

In 1991, the rival McLaren team took this development one stage further, using a very high-speed serial interface which allows data to be transferred from the onboard logger to a special computer in 10 seconds. McLaren would not reveal how much information this was capable of retaining, but it was believed to be adequate for an entire race.

Totally separate from the Williams team's system of chassis performance monitoring is the solid state recorder employed by the Renault Sport engineers, which facilitates not only the monitoring of all engine performance parameters, such as oil pressure and water temperature, exhaust gas temperature, fuel consumption and revs, but also allows engineers to focus in on engine performance through individual corners.

Because of the difficulties involved in sustaining a radio signal throughout a lap in connection with this information gathering process, the information is saved up, and when the car passes the pits a radio signal prompts the system to download all the information in a split second. Continuous telemetry throughout the lap, with gaps in the transmission caused by the car going behind buildings or grandstands would be of less value, and is fraught with more potential problems.

CHAPTER 8

INDY CAR TECHNOLOGY

While European-style Grand Prix racing is inevitably a technical package capable of tackling a wide variety of circuits from the outset, Indy car racing has grown to be the strong second force on the international single-seater stage as a result of the interest focused on a single specific race – the Indianapolis 500.

Grand Prix cars have had to display a versatility bred out of a requirement to perform on such disparate circuits as Monaco and Silverstone. The tremendous success of Indy car racing has been achieved very much on the back of that famous super speedway race, which was first held on Memorial Day 1911 and won by Ray Harroun at the then remarkable average speed of 75 mph (121 km/h). From then on, the single development of single-seaters in the US revolved round the high-speed banking in Indiana which fast became the jewel in America's racing crown.

Ironically, although much has been made of the European representation which has built up at Indianapolis over the past three decades in terms of car design and construction, in fact, this is in no way a recent trend. Moreover, the popular notion that Indy car racing was something of a technical backwater until the rear-engined revolution, inspired by British constructors in the mid-1960s, is simply not realistic. Ever since the 1920s, the business of Indy car racing has been every bit as technically innovative and specialized as European Grand Prix racing, despite the totally different nature of the prevailing competitive environment.

Although contemporary American technology produced the Marmom Wasp driven by Harroun to win the 1911 race, and the four-cylinder National – developing 120 bhp at 2,200 rpm – in which Joe Dawson triumphed the following year, the Europeans soon started to muscle in on this area of the sport.

In 1913 Peugeot arrived on the scene with what were probably the most significant and imaginative racing car designs of the immediate pre-First World War period. With their radical twin-overhead camshaft, four-cylinder engines featuring four valves per cylinder, pent-roof combustion chambers and integral cylinder head/block assembly, they would become the much-copied inspiration for a whole host of subsequent racing engines, including the historic Offenhauser unit. Mercedes would win in 1915 with Ralph de Palma at the wheel, and the huge 4.5 litre, four-cylinder engine used in his car featured fabricated cylinder block/head construction, also featuring four valves per cylinder, but with a single overhead camshaft configuration.

In fact, no American car would win the Indy 500 between 1913 and 1919, despite relatively competitive showings from the like of Stutz, Duesenberg, Miller, Frontenac and Maxwell.

The 1920s, however, would belong to US constructors with the American Automobile Association (AAA) adopting the same 3 litre capacity limit in 1920 that was being enforced for European Grand Prix racing. This sustained overseas interest into the 1920s, but although Ballot and Peugeot sent teams over to represent Europe, victory fell to the Americans, with cars owned and designed by Louis Chevrolet, with engines designed by the young engineer Cornelius Van Ranst, winning the 500 in 1920 and '21. Rounding off the 3 litre era, Jimmy Murphy triumphed in 1922 at the wheel of his Miller-engined Duesenberg hybrid.

The Duesenberg brothers, Fred and Augie, would be pivotal characters in the development of Indy cars through the 1920s, but probably the most significant development of that period was Harry Miller's discovery of two men who would help him shape the development of the American Championship car for more than the next decade – Leo Goossen and Fred Offenhauser.

By the early 1920s, most Indianapolis constructors were using eight-cylinder engines. Both Miller and Duesenberg followed the example of the earlier Peugeot unit, employing twin-overhead camshaft layout and a centrally mounted sparkplug for the best possible breathing and optimum combustion qualities.

Forced induction engines also reared their head during

this period, Duesenberg becoming the first manufacturer to install a centrifugal-type supercharger on an Indy racer as early as 1924. Miller, who also pioneered the use of front-wheel drive at the speedway, soon followed down the supercharged route.

In 1930 came another sea change. The effects of the Great Depression and the 1929 Wall Street crash threw concern over the spiralling costs of American Championship racing into sharp perspective, a historical detail which should remind those similarly concerned with the cost of Grand Prix racing in the early 1990s that there is Nothing New Under the Sun.

In an effort to contain cost, the exotic and specialized Indy cars of the 1920s gave way to machines built to a new formula which required a 6 litre (366 cu in) limit with a maximum of two valves per cylinder. Superchargers were banned on four-stroke engines, but that didn't prevent Leon Duray from coming up with a radical 16-cylinder supercharged two-stroke engine working a central eight-throw crankshaft. Perhaps predictably, its over-complexity ensured that it was not a success.

The most significant development in the 1930s, however, was the arrival on the scene of the four-cylinder Miller unit, an off-the-peg racing engine which was absolutely suited for the times, in the same way as the Cosworth-Ford DFV would slip into a similar snug-fitting niche just over 30 years later.

Miller's requirement was for a neat little unit which could be made available for less then $2,000 (then the sterling equivalent of around £500). It had to be suitable for either Indianapolis or the dirt tracks which comprised the backbone of the AAA schedule of the times. It had to be compact and light, reliable and capable of producing 150-200 bhp at around 5,000 rpm, running on gas/benzol fuel and survive a full season with the minimum of maintenance.

Leo Goossen duly obliged his boss with a superb technical product, although ironically Harry Miller seemed to lose interest in speedway racing after the 1930 supercharger ban. His business interests became less sharply focused, and having become bogged down in projects to develop economically unproductive sports car and aircraft engines, Miller eventually filed for bankruptcy early in 1933.

This was a crucial turning point in the history of Indy car racing. The development of the Goossen-designed four-cylinder engine could have ended there and then, but Fred Offenhauser scraped together all his savings and bought the engine tooling and parts inventory, thereafter setting up shop on his own.

In 1934, after a truly precarious financial period, he formed Offenhauser Engineering, re-hired Goossen and a couple of former Miller machinists – and never looked back. In 1935, Kelly Petillo won the 500 with the new, purpose-built Offy engine to entrench the company's presence firmly at the Speedway.

By the late 1930s and early 1940s, the character of the post-war Indy roadster was starting to take shape. In 1937 competition was further stimulated with regulations that again permitted the use of superchargers, and this triggered the design and construction of three totally new full race engines which were to have a deep influence on Indy car racing for many years thereafter. Needless to say, all three were laid out by Leo Goossen, for the reputation of Fred Offenhauser's company by now reached the point where anybody wanting a bespoke racing engine had only to outline its required specification, provide the necessary, and then approach Goossen and Offenhauser for it to be designed and manufactured.

The three engines concerned were the Sparks-Thorne six, the Meyer-Bowes straight eight, and the Winfield-Welch V8, later to become the famous Novi. The Novi, of course, would not experience its greatest days until after the Second World War, by which time Maserati had won two Indy 500s (1939 and '40), thanks to the efforts of Wilbur Shaw, while Floyd Davis and Mauri Rose co-drove the winning machine – essentially an uprated sprint car, so characteristic of the Indy chassis of the immediate pre-war era – in 1941, the final Memorial Day classic to be run until 1946.

That year also saw Louis Mayer and Dale Drake buying out Fred Offenhauser's business and hiring back Leo Goossen to take over the task of full-time design work. With significant financial backing, they believed that the best days of the Offy engine were still ahead. It was a correct judgement and, with astonishing improvements progressively incorporated into a design which had started out 15 years before, this team would continue offering admirable service to the Indy car fraternity for more than another 20 years.

The most impressive immediate pre-war car to arrive on the Indy scene was unquestionably a revised version of the Novi which appeared in 1946. It incorporated a new front-wheel-drive transmission designed by Leo Goossen and a chassis frame produced by an ambitious, enterprising young engineer from Los Angeles called Frank Kurtis. The Novis quickly proved themselves to be the fastest cars at the Speedway in those early post-war years, but they were desperately heavy on fuel and tyres. As a result, they were obliged to make more pitstops than their contemporary opposition, and consequently never bettered a best-ever third place achieved in 1948.

Frank Kurtis's chassis designs caught the imagination of the Indy car fraternity in the late 1940s and the Los Angeles engineer would spearhead a whole new generation of car builders, such as Ed Kuzma, Emil Deidt, Gordon Schroeder and, some while later, A. J. Watson.

Technical development continued apace in the hands of this enthusiastic group. Kurtis would experiment with a De Dion rear suspension at the speedway in at attempt to reduce rear unsprung weight in 1948. Car owner Eugene Cassaroll would initiate the development of telescopic shock absorbers with an approach to the Monroe company in 1959.

Disc brakes appeared, as would magnesium alloy wheels from Ted Halibrand's company, first to be used on the first Kurtis roadster – the Howard Keck Special – in 1952. And remember, all this was taking place at a time when the snooty, so-called high-tech world of Grand Prix racing was still grappling with drum brakes, spoked wire wheels and transverse leaf springs on the opposite side of the Atlantic!

Lou Meyer and Dale Drake, meanwhile, were encouraging Leo Goossen to initiate changes and improvements to the Offenhauser engine – even though it faced no real competition in the world of big-time American track racing. One of the most significant developments was the introduction of fuel injection in the late 1940s, a major performance breakthrough pioneered on Californian hotrodder Stuart Hilborn.

Initially the demand was for bolt-on systems to be used on Offy-engined midget racers, then enjoying a peak of popularity on the west coast of the USA, and it was tried at Indy on the Howard Keck Special as early as 1949.

Even diesel-engined cars had their moment of glory at the Brickyard. In 1952, Fred Agabashian qualified the sensational Cummins Diesel on pole at a record 139 mph, its participation prompted by the AAA permitting diesel engines with a capacity limit of 402 cu in from 1950, with or without superchargers.

In many ways, the Cummins diesel represented a turning point of Indy car chassis design as well. Its Frank Kurtis-designed chassis offered a low, ground-hugging profile. Its engine was canted over to the right, thereby positioning the crankshaft centre line to the far left of the chassis, enabling the propeller shaft to run to the left of the cockpit which allowed a very low seating position.

A quick look at the corner speeds achieved in 1952 by Agabashian's mount and the other Kurtis roadster – the Howard Keck Special driven by Bill Vukovich – and it was clear that the old sprint-type car, with its high centre of gravity, was a dead duck at Indianapolis. It would survive on the dirt tracks, sure enough, but its days as a competitive tool at Indiana's banked speedway were surely over.

From 1953 to '62 the Indy 500 would be dominated by the legendary roadster. At the sharp end of this particular technology stream, Kurtis would now be joined by Kuzma, A. J. Watson and Quinn Epperly. Watson exploded on to the scene quite suddenly in 1956 when his first design scored a debut win in the 500 in the hands of Pat Flaherty, and his cars would soon become a part of Indianapolis folklore.

Eschewing the Cummins diesel-style canted engine installation, Watson mounted his Offy engines vertically in the chassis frames – but positioning both engine and drive train far further to the left than had ever previously been tried. Epperly, meanwhile, worked in collaboration with car owner George Salih to introduce canted engine installation which Sam Hanks would use to win the 1957

The way we were. Bill Cheesbourg's Novi-engined Kurtis roadster which ran in the 1958 Indianapolis 500 was typical of the front-engined Indy car of that decade.

500 in the Belond AP Special, an achievement duplicated by Jimmy Bryan the following year.

Ironically, while the roadsters continued their reign at Indianapolis, it would be a chance remark by a member of the Indy car establishment which resulted in the technical complexion of the business being upended in the early 1960s. The man responsible was Rodger Ward, winner of the 500 at the wheel of a classic Watson-Offy roadster in both 1959 and '62. During the interval between his two victorious outings, he was responsible for triggering a chain of events which sounded the death knell for the very cars he had driven into the history books.

Fascinated by the world of Formula 1, Ward contested the first 1959 United States Grand Prix at Sebring in an Offenhauser-engined midget car. Before practice, he advised the works Cooper team that it was wasting its time, believing that his home-brewed machine would easily out-corner their state of the art, contemporary, rear-engined F1 machines.

Ward was soon disabused of that notion, and quickly became a convert to the rear-engined philosophy. In fact, the Indy winner told John Cooper straight that a car such as their F1 machine could turn the Indy establishment on its ear. Cooper heeded his words, and in October 1961 one of the team's F1 Cooper-Climax T51s was shipped to the speedway where Jack Brabham duly gave it its first test.

Brabham, twice World Champion, nevertheless had to undergo the official Indianapolis 'rookie test' which involved an eight-lap acclimatization phase, during which lap speeds were not supposed to exceed the 115 mph (185 km/h) mark. Jack was then allowed to skip the 120 mph (193 km/h) phase, and was then flagged off by an irate Harlan Fengler, the Indy Chief Steward, and given a chewing out for going too fast too soon, when he posted a lap at 128 mph (206 km/h)!

By the end of the test, the spindly little Cooper-Climax had returned a best lap at 144.83 mph (223.03 km/h) – a speed which would have earned Brabham eighth place on the grid for the 1960 Indy 500. Perhaps in recognition of identifying the Cooper's potential in the first place, Rodger Ward was also invited to try a few laps in the car, and he returned to the pits bubbling with enthusiasm for its smooth ride and outstanding stability. The only pity, he mused, was that it hadn't got a little more steam...

Thus encouraged, Cooper and Brabham made plans to return to contest the Indy 500 in 1961. With sponsorship

A. J. Foyt was one driver whose Indy car career extended through the years of momentous technical change on the US racing scene. He would run at the 1992 Indy 500 at the wheel of a state of the art Lola, but his first outing at the Brickyard had been 34 years earlier in this Clint Brawner-prepared roadster.

from Kleenex tissue heir Jim Kimberley, and despite lack of familiarity with the oval racing technique, Jack brought the Climax-engined special home ninth in a race which had been dominated by the Watson-Offy roadsters of A. J. Foyt and Eddie Sachs. Foyt's winning average was 145.903 mph (234.758 km/h), the little Cooper slogging home ninth at 134.116 mph (215.793 km/h). It was a performance which set a time bomb ticking away under the front-engined Indy roadsters. The Brickyard would never be the same again.

In 1962, American Dan Gurney was destined to utter words every bit as prophetic as those which issued forth from Rodger Ward three years earlier. Viewing Colin Chapman's F1 Lotus-Climax 25 on its race debut in the pit lane at Zandvoort, the lanky Californian reportedly exclaimed, 'My God, if somebody took a car like that to Indianapolis, they could win with it!'

Gurney invited Chapman to watch the 1962 500, and the Lotus boss could not believe his eyes.

'I thought I'd gone back 15 years. I could imagine this was what it must have been like to watch the Mercedes and Auto Unions pre-war at Tripoli. I thought, well, all you've got to do is to get an engine with about half the power of these great lumps of junk, build a decent chassis and you've won the race!'

A lofty assumption indeed; but Chapman was about right!

At about the same time that the Lotus boss was mulling these thoughts over in his mind, so the Ford Motor Company was also pondering the question of Indianapolis from its corporate headquarters in Detroit. Ford engineers Don Frey and Dave Evans had also attended the 1962 Indy 500, but at the same time as Evans began making approaches to leading USAC chassis men, Chapman was also scheming an approach to the US auto giant.

Within a couple more months, Chapman had completed his planning. He worked out a scheme whereby what amounted to a low-powered 'economy car' could complete the 500 with only one fuel stop rather than the two or three which were usual for the front-engined roadsters.

The deal was eventually done, clinched by Jim Clark's remarkable pace at the Speedway with the 1962 US F1 Grand Prix-winning Lotus 25. A purpose-built machine, the Lotus 29, was constructed to take a 4.2 litre race-tuned version of the Ford Fairlane running on regular gasoline rather than alcohol. With Clark at the wheel, a four-lap average of 149.750 mph (240.947 km/h) placed him fifth on the grid, with Gurney earning a fourth row start in the second entry at 149.019 mph (239.772 km/h).

The story of the 1963 Indy 500 has gone down in the history books as a battle not so much between old and new technology, but more as a tussle between the speedway establishment and a bunch of precocious newcomers. In this case, the newcomers were obviously Chapman, Clark and their Lotus 'funny car', while the establishment

Parnelli Jones's Indy win in 1963 with J. C. Agajanian's Watson-Offy roadster came in the face of a strong challenge from the compact Lotus-Ford driven by Jim Clark – and was achieved with a dash of complicity from the Indy establishment!

was represented by Indy car owner J. C. Agajanian, his driver Parnelli Jones, and their famous Watson-Offy roadster which started the race from pole position.

In the early stages of this historic race, Jones broke record after record, making his first routine fuel stop after 64 of the 200 laps, dropping back briefly to eighth place and allowing Roger McCluskey's similar Watson-Offy to take the lead.

Shortly afterwards it began to dawn on the Indy establishment just what a formidable challenge was being offered to the established order by the Lotus-Fords. It was now clear that they would be able to run through to the finish on two fewer pit stops than their more experienced rivals, and by lap 80 Clark and Gurney were running first and second, with Parnelli in third place but charging hard to make up lost ground.

Eventually Jones made it back into the lead, but there was just no way he could shake off the bottle-green Lotus. Then the leading Watson-Offy began laying a trail of lubricant, apparently from a horizontal split in its rear-mounted oil tank. On the start line, Chief Steward Harlan Fengler was urged by Agajanian not to black flag him until he had studied the problem through field glasses. Agajanian would later remember the episode:

'I had to stop him somehow. I said "We were throwing oil, but we're not any more. Harlan, you can't call us in now." And just then Colin Chapman comes rushing up with his English accent and says, "Pull that car off there before he kills all the other drivers!" and I yelled at him, "Get your butt back over that wall, Chapman... Fengler and I are talking. It's not your car that's in trouble." And Fengler turned to Chapman and said, "Get back please, Mr Chapman..." and all the while I was saying you can't take the race away from us this late, you just can't do it.'

Fengler agonized, studied Parnelli's car through the glasses for another two or three laps and then nodded, 'OK, Aggie, we'll let him finish...' The establishment had won, and Chapman took a long time to stop bristling about the way in which he had been treated. Clark finished second behind Jones.

For the '63 race Ford had managed to coax 350 bhp on gasoline and carburetters out of the Lotus 29s, and with 46 gallons of fuel tankage available, these monocoque machines tipped the scales at 1,300 lb (590 kg) dry, a good 300 lb (136 kg) lighter than the most competitive roadster.

When Colin Chapman and Jim Clark brought the F1-derived Lotus 29 to Indianapolis in 1963, it instantly – and irrevocably – rewrote the parameters of Indy car design. Clark finished second behind Parnelli Jones on Lotus's maiden outing at the Brickyard.

In 1964 it was Detroit's intention to increase the power output by 50 bhp to over the 400 bhp mark, an ambition realized by incorporating a variety of improvements including twin-overhead camshaft heads. However, it was clear that carburation would not be adequate for the increased airflow demands of the four-cam engines, so a switch to Hillborn constant-flow fuel injection was the only choice.

Ford engineers had not been particularly anxious to adopt this route as this method of fuel metering was not

particularly precise over a broad engine speed and load range. This had not posed too much of a problem with alcohol fuels, as engines would tolerate broad deviations from the optimum fuel/air ratio without a significant loss of power. However, gasoline was far more critical, but Ford was determined to run with gasoline fuel for at least the first race with its new four-cam V8.

For the 1964 race Ford also supplied engines to A. J. Watson, to power his new rear-engined creation which would be driven by Rodger Ward, for the Halibrand Red Ball Shrike driven by Eddie Sachs and the ex-Clark Lotus 29 driven under the Lindsay Hopkins team banner by brilliant rising star Bobby Marshman. In the opening stages of the race, Clark and Marshman ran away from the pack, but both retired and A. J. Foyt was allowed through to score the last ever victory for the classic front-engined Watson-Offy roadster.

In 1965 Clark returned to win for Lotus, followed by Graham Hill's success the following year in a John Mecom Lola. British rear-engined technology effectively wiped out the traditional roadster within two years, but no sooner had these new technical parameters been established than the Indy fraternity found itself the focal point for yet more technical turmoil as a frenzied period of development on four-wheel drive systems and turbine engines made its mark on the Speedway's history.

In the years following the final appearance of Harry Miller's last four-wheel drive design in 1948, Indianapolis had settled down into a predictable pattern which was only disrupted by the arrival on the scene of Jack Brabham's Cooper-Climax some 13 years later.

Four-wheel drive then made a return to Indianapolis in 1964 after Stirling Moss met colourful STP boss Andy Granatelli, and enthusiastically recounted how the Ferguson four-wheel drive system had served him so well in the 1961 Oulton Park Gold Cup. Granatelli absorbed what Moss was saying and subsequently asked Tony Rolt, the boss of FF Developments, whether Ferguson would send the P99 car over for trials at the Brickyard.

Rolt agreed, and Jack Fairman recorded some consistent laps with the 2.5 litre machine at around the 140 mph (225 km/h) mark, leaving Granatelli highly impressed and ready with a proposition that would virtually require Ferguson to design him a new Indycar.

Conceived and constructed in five months at Ferguson's Coventry base, the result was a front-engined chassis of spaceframe construction with a semi-monocoque centre

A. J. Foyt heading for the final front-engined victory that would ever be scored at Indianapolis, at the wheel of his Watson-Offy roadster, in the 1964 500 – only four years later than the last front-engined car won a World Championship Grand Prix.

section. Power came from one of the famous 268 cu in Novi V8 engines developing about 740 bhp, the power being transmitted via a four-speed constant mesh gearbox to a four-wheel drive system broadly similar to that used on the P99.

This had obviously been scaled up to deal with the extra power and had a torque split variable from 70/30 to 60/40, whereas the P99's split was a finite 50/50. Entered for Jim McElreath to drive in the 1964 Indy 500, it was subsequently taken over by Bobby Unser, who unfortunately tangled with another car while trying to avoid the fatal accident to David MacDonald and Eddie Sachs on the second lap. The race was stopped, and although Unser was not injured, it proved impossible to repair the Novi Ferguson in time for the restart.

Unser came back in 1965 with the same car, qualifying at 157.467 mph for a start in the middle of the third row, and was running seventh after only 10 laps. He subsequently fell away and retired with an oil leak after 180 miles' racing, but the seeds of enthusiasm for serious four-wheel drive systems at Indianapolis had been sown.

After a year's absence from the 500, four-wheel drive made a serious return to the grid in 1967. But it wasn't this aspect of Andy Granatelli's new car that grabbed everybody's attention, for this was the striking new STP

Bobby Unser in the Ferguson-Novi 4WD roadster in the 1964 Indy 500. Four-wheel-drive systems would be applied with more promise on the rear-engined Indy cars later that decade.

Turbine car, variously dubbed 'Silent Sam' and the 'Swooshmobile'. With Parnelli Jones at the wheel, it stroked away from the opposition, and despite a spin was dominating the race, only for ballbearings in the transmission to fail with a mere seven miles to run before the chequered flag.

Arguably the most revolutionary and almost certainly the most distinctive car ever seen at the Brickyard, this

The Wood brothers' legendary NASCAR refuelling team was hired especially to service Jim Clark's Lotus-Ford V8 at Indianapolis in 1965 when he won commandingly. The off-set on the front suspension can clearly be seen in this shot.

ENCO
STP
OIL TREATMENT
40
STP
40

31
32
33
STP
STP
STP
AMERICAN
STP
Firestone

turbine car was the product of Granatelli's links with the Paxton division of the Studebaker Corporation. They modified the Ferguson transmission system, while the Montreal-based Pratt and Whitney organization produced a turbine motor turning out over 500 bhp at 1,350 lb pressure. The monocoque backbone chassis had the engine slung along one side and the driver along the other, and it was the Paxton system of transfer gears from the turbine which proved to be the root of the eventual trouble. Despite remonstrations from the other teams, the turbines seemed set to stay.

Colin Chapman watched all this development with interest. For 1968 he forged a deal with Granatelli under which STP sponsored his distinctive, wedge-shaped Lotus 56, which also relied on Pratt & Whitney motors for its power. Substantially lighter than the STP Paxton car, the engine in the Lotus was offset to the right in order to accommodate the drive-train on the left.

Disappointingly, despite their electrifying speed, the Lotus 56s were to be surrounded by tragedy. Jim Clark came briefly to test one at Indy shortly before his death in an inconsequential Formula 2 race at Hockenheim. A distraught Colin Chapman found BRM's Grand Prix number one, Mike Spence, willing and contractually able to drive the car, but he crashed heavily during testing and died as a result of head injuries.

Nevertheless, by the time the race had started, both Joe Leonard and Graham Hill occupied the first two places on the grid, and the third Lotus 56, driven by Art Pollard, had qualified on row four.

Again, the turbines looked extremely impressive, but again victory would slip from their grasp in the closing stages. Leonard was leading with nine laps to go when he ground to a halt. Pollard had gone out three laps before. The yellow light had been on for a number of laps in the wake of an accident, so the turbines were running along at reduced power. When the green light went on, both the Lotus drivers floored their throttles hard, and the sudden extra load imposed broke the extension shaft in the fuel pumps of both cars.

The turbines were not the only serious entries in 1968 to be fitted with four-wheel drive. Hewland Engineering, the British specialist gearbox company, prepared a four-wheel drive system for the George Bignotti-owned Lola T150 entered for Al Unser. Owing much to Ferguson principles, the Hewland system was simple in concept, utilizing many components from other trans-

Left *The Indianapolis fraternity dallied with turbine engines during the late 1960s and Parnelli Jones, seen here during a pitstop with the STP turbocar in 1967, very nearly won the classic race on this occasion. STP boss Andy Granatelli is holding a lightweight aluminium funnel ducting the turbine engine's exhaust away from the mechanics who are refuelling the car.*

Left *From left to right: Granatelli, Jim Clark and Parnelli Jones with the high-tech Colin Chapman-designed Lotus 56 which featured ultra-clean aerodynamics, four-wheel drive and a turbine engine. Clark was killed in an F2 race at Hockenheim before he could race the machine in the '68 Indy 500.*

Right *Colin Chapman's most technologically advanced Indy car was the 1969 type 64, equipped with a turbocharged Ford V8 engine and a 4WD system. Andretti was lucky to escape a fiery accident in practice caused by a wheel hub problem. The car was withdrawn and Mario won the race in the Clint Brawner-prepared Hawk, a Brabham Indy car look-alike.*

missions in the Hewland range.

A DG main casing, crownwheel and gear cluster were used, but a new rear-end casing was fabricated containing a train of gears which side-stepped the drive out to the right. From there a propshaft ran to an offset LG differential at the front of the car.

Mike Hewland originally anticipated that Bignotti would be using a four-cam Ford V8, so it was with considerable misgivings that he heard news that a 700-plus bhp turbocharged engine was about to be installed in its place. But his doubts about the strength of his transmission system were unfounded, Unser qualifying sixth, only to spin disappointingly out of the race after only 40 laps.

Hewland's system could also be adapted for two-wheel drive, a fact which anticipated the actions of the United States Automobile Club over the next few years. There was considerable pressure building up from those who didn't like the idea of a turbine car, for many resented this intrusion into their staid and conservative world.

After the 1966 race, when it became apparent that turbines were on the way in, USAC attempted to head them off by restricting the turbine inlets to an area of 23.99 sq.in, then to 15.399 sq.in after Jones nearly won in 1967.

Granatelli immediately embarked on a gigantic legal action which took his dispute with USAC to the High Courts, but the writing was on the wall, and Chapman, for one, didn't bother about turbines for 1969.

Instead, the Lotus squad returned in association with STP using Ford turbocharged machines modelled much on the lines of the Lotus 56s, only considerably slimmer. These were the fearsome and doomed Lotus 64s, the first outing for which saw Mario Andretti slam into the retaining wall at 150 mph (241 lm/h) when a rear hub failed, owing to incorrect heat treatment, and a wheel fell off.

Chapman was unable to get redesigned hubs ready in time for Graham Hill and Jochen Rindt, the latter now showing the same lukewarm approach to Indianapolis that he would later demonstrate towards the four-wheel drive Formula 1 Lotus. Both cars were withdrawn and an enormous row blew up between Granatelli and Chapman, although Andretti went out and won the race in the two-wheel drive STP Hawk – a Clint Brawner-built Brabham lookalike – which gave him great consolation.

The four-wheel drive Lolas fared much more successfully. Bobby Unser put his Offy-powered example on the outside of the front row, this machine now updated to

IN
STP
2
STP OIL TREATMENT
Firestone

2 STP OIL TREATMENT
STP
OIL TREATMENT
2
Firestone
AMERICAN

Lola makes its mark. The 1969 type T150 featured a 4WD system developed by the British F1 transmission specialists Hewland Engineering, and was driven in that year's Indy 500 by Al Unser, father of the 1992 race-winner.

T152 specification, using Hewland transmission basically the same as that employed in 1968, although now with four speeds. Mark Donohue, who had a choice of two 152s (one fitted with a stock-block Chevy and the other with a turbo Offenhauser) started the latter car from fourth place on the grid.

As things transpired, this proved to be the pinnacle of four-wheel drive success at the Brickyard. Unser finished third behind Andretti and Dan Gurney's Eagle, while Donohue was seventh. That the Lola concept was successful is emphasised by the fact that George Bignotti prepared a similar, but two-wheel drive car for 1970–71, dubbed the Colt, which won the race both years in Al Unser's hands.

Now disputes between USAC and those interested in four-wheel drive and turbines flared up once again. After the 1968 race, USAC rejected a recommendation from its rules committee that an outright ban be placed on the turbines, even though they had been extensively limited by reducing the area of the intake annulus. But the real crunch came for 1970. After the 1969 race, four-wheel drive was banned at Indy for good, even though it had been allowed that very year, ostensibly to permit those who wanted to develop two-wheel drive machines some continuity of research.

The reasons for the ban were that four-wheel drive was not considered to be within the mainstream of automotive development, a statement which brought forth a vehement reaction from Ferguson in particular. They accused USAC of trying to stifle automotive progress, and totally dismissed their suggestion that four-wheel drive was never likely to be used on production road cars. Granatelli also waded in to point out that USAC was merely banning four-wheel drive for its success rather than any other reason, using it as a front for effectively banning the turbines for good. But USAC had the last word and that was that.

Into the 1970s it was McLaren who would emerge as one of the most potent and technically innovative forces on the Indy car scene, tackling a 500 challenge for the first time in 1970 with the Offy-engined M15. The US track racing establishment was highly impressed with the finish and preparation of the new cars, but their competitive edge was blunted when they lost their two top drivers before the race. Chris Amon was honest enough to admit that he simply couldn't get used to running with a brick wall a couple of feet from his right ear, while poor Denny Hulme received serious burns to his hands in a fire during practice, caused after a fuel filler cap popped open and allowed a haze of methanol to be ignited by the red-hot turbocharger casing.

The superb aerodynamic McLaren M16 brought F1 standards of aerodynamic detailing to Indy car racing in 1971. This shot of Peter Revson's Offy-engined machine emphasizes its elegant simplicity of line.

McLaren built the chassis to beat at Indianapolis during the early 1980s. Here Johnny Rutherford's works M16C pauses for a routine pitstop en route *to winning the 1974 500.*

Opposite *The F1-derived McLaren M24-Cosworth DFX's of Johnny Rutherford (works) and Tom Sneva (Penske) were highly competitive challengers throughout the 1977 Indy car season. The cars were closely modelled on the Gordon Coppuck-designed M23s which had won McLaren the World Championship in 1974 and '76.*

A few weeks afterwards, Bruce McLaren himself was killed at the wheel of a CanAm M8F whilst testing at Goodwood, but the team would bounce back to tackle Indy car racing with renewed vigour in 1971, when the Gordon Coppuck-designed M16 set the pace in terms of aerodynamic development at the Brickyard.

Supported by Roger Penske's private entry for Mark Donohue, the works McLaren M16s set the pace from the outset. Donohue dominated the '71 race until his gearbox broke, but returned the following year to nail a first win for the McLaren marque at the Brickyard. The M16 theme would be constantly developed over the next few years and Johnny Rutherford provided McLaren with Indy 500 victories in 1974 and '76, after which the historic machine gave way to the new F1 derived M24.

This was another classic example of Formula 1 technology being transferred to the Indy car environment. The McLaren M24 was closely derived from the Coppuck-designed M23 which had been introduced on to the Grand Prix scene at the start of 1973, and won the World Championship in the hands of Emerson Fittipaldi (1974) and James Hunt (1976). The new Indy car combined M23 engine mounts, an almost identical chassis layout built with heavier monocoque skins and bulkheads, and a sus-

FIRST NATIONAL CITY
2
McLaren
GOODYEAR

CAM2
SNEVA
MOTOR OIL
NORTON
8
GOODYEAR
SPIRIT
DieHard

The sensational Chaparral 2K was the first Indy car to have full F1-style ground-effect aerodynamics. Designed by John Barnard, it led the '79 Indy 500 in the hands of Al Unser before encountering gearbox trouble and then returned to win the following year with Johnny Rutherford at the wheel.

pension set-up derived directly from the McLaren M16E.

More important still, the McLaren M24 marked the Indy debut of the turbocharged 2.65 litre Cosworth DFX V8, a totally new engine derived from the 3 litre Cosworth-Ford F1 V8. This formidable combination made its Indy debut in 1977, since when Cosworth has not been absent from the F1 scene.

By 1981, McLaren's Indy star would have waned, Penske having set up as a constructor in his own right, while the advent of Formula 1 ground-effect aerodynamics opened a whole new avenue of development for Indy car constructors. In this respect, the most significant machine would be the superb John Barnard-designed Chaparral 2K which made its Indy debut – and nearly won – in 1979, before Johnny Rutherford used it to post his third career victory at the Brickyard the following year.

The 1980 race was also distinguished by the appearance of Bobby Hillin's Longhorn, a Indy racing version of the superb Patrick Head-designed ground-effect Williams FW07, which was busy rewriting the standards of Grand Prix car performance on the opposite side of the Atlantic. Disappointingly, it failed to realize its undoubted potential, but confirmed that ground-effect aerodynamics were the way to go for future Indy car concepts.

CART is born

The biggest sea change in Indy car racing, of course, was destined to take place off the track in the late 1970s. More controversy was brewing behind the scenes. Throughout

Al Unser at the wheel of the Longhorn-Cosworth at the 1980 Indy 500, this being, in effect, a ground-effect Williams FW07 chassis uprated to conform with the Indy car regulations. Although the project had a lot of potential, it lacked development and consequently went short in terms of hard success.

that decade, USAC had been attempting to restrict speeds with a combination of technical initiatives, including reductions in fuel allowances and capacity, and boost pressure limitation valves.

In 1972 the first fuel restriction since the 1930s was imposed at Indianapolis, although it was set at 325 gallons per car and thus caused few problems. This was progressively reduced, with a 280 gallon allotment announced in 1974, along with an 80 in manifold pressure limitation.

From that point onwards the pit lane at Indianapolis became a hive of activity as teams attempted to devise methods of circumventing the relief valves, mainly by over-boosting to the point that the valve just couldn't cope. Finally, when USAC slashed boost limits for Cosworth DFX users from 80 to 50 in of manifold pressure for 1979, all hell broke loose when many teams adopted the technique of over-boosting by interfering with the performance of the wastegate valves. The biggest problem now facing USAC was to ascertain which teams had been cheating during their qualifying runs and which teams hadn't.

In the wake of this drama, legal actions were threatened from all sides. USAC weathered the storm, but the controversy gave added impetus to the aspirations of Championship Auto Racing Teams Inc (CART), which had been co-founded by team owners Roger Penske and Pat Patrick the previous year.

Penske and Patrick represented the tip of the iceberg of disillusionment with the way in which USAC was administering Indy car racing. Along with their fellow team owners, Penske and Patrick proposed an increased level of car owner participation on the rule making and administrative aspects of the sport.

Initially the 21-man USAC Board of Directors vetoed the idea unanimously, so the fledgeling CART organization announced that it would conduct its own series of races, and on 30 November 1978 it elected Patrick as President and Jim Melvin as Vice-President and General Manager.

The foundation of CART was based on a number of fundamental concepts, including an increased responsiveness to the competitors, sponsors, promoters, news media and fans. It set out to enhance the economic viability of the series with the intention of stabilizing rules and improving the overall image of Indy car racing.

CART staged its first race on 11 March, 1979 when Gordon Johncock won a 150 mile event at the Phoenix International Raceway mile oval. But a month later, USAC launched a counter-attack, its board of management voting to reject the Indianapolis 500 entries from Penske Racing, Patrick Racing, Chaparral Racing, Team McLaren, Fletcher Racing and Dan Gurney's All American Racers. Lawyer John Fracso, acting on CART's behalf, filed suit to have the teams reinstated. On 5 May, barely three weeks prior to the Memorial Day Classic, District Judge James F. Noland decided in CART's favour, obliging USAC to lift its ban.

Eventually CART and USAC would learn to live together. There was no other choice. CART sanctioned the vast majority of the races, but not Indianapolis. USAC was effectively left marooned, yet presiding over the jewel in the US single-seater racing crown. All the other races effectively acted as previews or postscripts to the Indy 500, although this polarization of attention around a single event would gradually become less pronounced throughout the 1980s, as the whole Indy car series developed a rounded appeal in its own right.

CHAPTER 9

INDY CAR CHASSIS DEVELOPMENT

After Colin Chapman's Team Lotus struck what amounted to technical gold with the development of ground-effect aerodynamics in Formula 1 during the mid-1970s, it was only a matter of time before the technology was transferred to the world of Indy car racing. The true Indy-type ground-effect racers were seen at Indianapolis in 1979, in the form of the aerodynamically innovative Penske PC7 and the Chaparral 2K, the latter alone maximizing of the under-car aerodynamic possibilities with full-length 'tunnels' extending back beyond the side pods to the very rear of the car.

The Chaparral was the work of John Barnard, formerly Gordon Coppuck's design assistant at McLaren, and would be a car crucial to establishing the Englishman's reputation as one of the most imaginative and original-thinking engineers on the international motor racing scene during the 1980s.

Barnard had worked with Coppuck on the M23 and also became involved with revisions to the successful McLaren M16 Indy car design. He remembers:

'On the later M16 versions I became much more closely involved in the business of the chassis design. I completely redrew the tub front to back for the M16E, because over its long development life all sorts of nasties had grown inside to support various modifications. It had increased in weight and complication, so it was time to re-do the design, save weight and stiffen it all up. We kept the outer monocoque skins, but the innards were all revised.'

In 1975 Barnard joined the Vel's Parnelli Indy car team,

based in Torrance, California, working briefly on the team's Lotus 72-inspired F1 car which had originally been designed by Maurice Phillipe, before producing the highly successful VPJ4B Indy car for the team, which eventually collapsed and withdrew from racing in 1977.

At the end of 1977, Chaparral boss Jim Hall approached Barnard and asked if he would design an all-new Indy car for his Texas-based team. John agreed, but returned to England to design and build the machine. It was eventually built by Bob Sparshott's Luton-based BS Fabrications company, which had run US privateer Brett Lunger's McLaren M23 and M26 F1 cars in 1977–8.

With Al Unser at the wheel, the full 'sliding skirt' Chaparral 2K led the '79 Indy 500 commandingly until its gearbox failed, and then came back to win the race the following year. By then, however, Barnard had quit the project, fed up with the amount of adulation the media focused on 'Jim Hall's brilliant new design', which the British engineer believed seriously short-changed his own contribution to the project at a crucial time when he was building his career.

Barnard would also play an indirect part in raising the standards of Indy car chassis technology on another front when March Engines, the Oxford-based special projects off-shoot of the prolific Bicester-based constructor, was approached to produce a ground-effect chassis for long-time Indy car entrant Sherman Armstrong.

Armstrong made it a condition of the deal that Barnard should play a major role in the engineering input, but by then the British designer was so heavily involved with Ron Dennis in the formation of the McLaren International organization that there was no way in which he could play an active role in the programme.

What he did manage to do, however, was to set out a few basic parameters which provided guidance to March Engines on such matters as the appropriate aerodynamic package. The resultant car, dubbed the Orbiter, was built around an F2 March 792 chassis and used the ubiquitous Cosworth DFX engine. It certainly displayed some worthwhile potential, although its driver, Howdy Holmes, failed even to complete a qualifying run at Indianapolis.

This programme sowed the seeds of the March company's Indy car involvement, a programme which would raise the standards of competition even further throughout the 1980s. After Gary Bettenhausen finished third at Phoenix with the Orbiter in the final race of the year, behind Ton Sneva's Phoenix and Mario Andretti's Wild-

Carbon fibre chassis construction was allowed only sparingly on the Indy car scene in the early 1980s. (**Above**) *Tom Sneva poses in his Mayer Motor Racing March 84C in 1984, showing the carbon fibre upper section which was then permitted by the rules, while the late Al Holbert* (**left**) *stands by the March-built 1987 Porsche Indy car which shows very distinctly the dividing line between the carbon fibre upper section of the monocoque and the aluminium honeycomb lower section.*

cat, March would receive an approach from respected Indy car operator George Bignotti.

He proposed that March should design four complete corners for his own upcoming Indy car project, but March boss Robin Herd persuaded Bignotti that the British company should build him a complete car. The well-heeled Florida-based Don Whittington then also turned up on the doorstep to express interest, and ordered cars to be raced by himself and his brother Bill, and before long Herd found himself pledged to deliver several March Indy cars in time for the 1981 500.

On the face of it, Herd's assurances seemed wildly optimistic. On the other hand, Indy car racing was in some ways a relatively low-tech affair by the F1 standards of the time, and mindful of how profitable it had been for Williams when Patrick Head's FW07 design was metamorphosed into the Longhorn for oil man Bobby Hillin's team, Herd quickly came to appreciate the potential offered by this new market. His own RAM March 811 owed its inspiration to the Williams FW07, so it was now reworked to form the first March Indy car.

Its structure strengthened with a thicker gauge outer monocoque skin to conform to the Indy car regulations, the resultant machine retained almost identical inboard coil spring/dampers, activated by fabricated rocker arms, the pivot points at the front being well out in the airstream. The car also featured a transverse Weissmann gearbox design, enabling the rear underbody airflow to be cleaner than that of any existing rival Indy car designs.

The 1981 Indy car season saw ground-effect technology exploding across the board with 27 of the 33 runners in that year's 500 displaying this aerodynamic configuration. Thanks to the entrepreneurial efforts of the CART organization, Indy car racing was now enjoying better support and promotion, attracting the biggest crowds for more than a decade and more television than ever. A previously moribund series under the unimaginative and conservative control of USAC, it was now dramatically revitalized, and it was against this upbeat, optimistic backdrop that March Engineering made its Indy car debut.

It was predictable, however, that Indy car racing would suffer what amounted to a re-run of the F1 controversy over aerodynamic side skirts, CART regulations permitting them to make contact with the ground, while USAC stipulated a ban on any skirt which extended below the bottom of the car's chassis. Despite that, and the 48 in boost restriction now imposed at the Brickyard, the qualifying times for the 1981 Indy 500 proved quite remarkable, with the new March 81C, fresh out of the box, setting the pace during the second qualifying weekend.

Pole position went to Bobby Unser's Penske PC9B at 200.545 mph (322.678 km/h), but had it not been for the late delivery of the first Bignotti-Cotter team March, Tom Sneva might well have been right at the front. Bignotti had actually managed to get the car out on to the circuit at the end of the first practice week, but some gearbox damage, caused by the starter pinion repeatedly jamming in the ring gear, forced the team to return to its base to resolve the problem.

Come the race itself, Sneva proved every bit as much a threat as his qualifying performance suggested. At the start, Bobby Unser surged into the lead from Johnny Rutherford's Chaparral 2K, A. J. Foyt's Coyote and the Longhorn 'FW07B' of Al Unser. But by lap 24 Sneva had surged into the lead, and remained a force to be reckoned with before pulling off after 97 laps with gearbox failure.

By the end of the 1981 season March was firmly entrenched as a front-line Indy car constructor. Gordon Kirby, respected American correspondent for the well-regarded British Magazine *Autosport*, summed it up thus:

> 'The big point about the March 81C was that it had downforce at a time when Indy car technology was in the Dark Ages from the aerodynamic standpoint. As an example, two years earlier, the Penske PC7 had set new standards, straight out of the box. It established bench-marks that the PC9s were struggling to match several seasons later. In that respect, March timed its Indy car debut perfectly.'

In 1982, March did its own Indy car transmission for the first time, the first example of which was flown out to the first race at Phoenix and fitted to Pancho Carter's 82C, which finished in sixth place. But the Penske PC10 proved to be the car to beat on the Indy car scene and the March efforts at Indy were over-shadowed by the horrifying fatal accident which befell Gordon Smiley, whose 82C was atomized in a turn-three head-on qualifying accident. It was the first fatality at the speedway since Swede Savage died as a result of injuries sustained when he crashed his Eagle in the 500 nine years before.

Smiley's death, along with that of rookie Jim Hickman at Milwaukee later in the year, triggered efforts to impose some sort of aerodynamic restrictions on Indy car performance for the following season, in much the same way as banning ground-effect underbodies was intended to reduce cornering forces in F1.

However, rather than make sudden across-the-board change to flat-bottomed chassis, as F1 did at the start of the same year, CART and USAC bore in mind both driver safety and the nagging question of chassis obsolescence by raising the bodywork/skirts a full inch above the chassis baseline while moving the rear wings forward by 7 in (18 cm).

Commented Al Unser at the time, 'This has reduced the ground-effect to the point where the drivers' reactions can keep up with what the car is doing – and also means that the driver has got to apply his abilities to greater use.'

However, preliminary tests of the first March 83C – held at Indianapolis in November 1982 – made the prospects of the new regulations putting the lid on the lap speeds looking something of a forlorn hope.

With totally revised aerodynamics finalized by Ralph Bellamy working in the wind tunnel at London's Imperial College, the 83C, with its one-piece carbon fibre undertray, was touted by March as producing a break-through similar to that achieved by ground-effect.

Robin Herd had outlined lower weight, increased chassis stiffness and more efficient aerodynamics as the funda-

Below and right *Mario Andretti came close to winning the 1985 Indy 500 in this Nigel Bennett-designed Lola T800, the chassis design of which benefited considerably from Bennett's previous F1 experience and started Lola along a path which would eventually end the early 1980s domination of March Engineering on the Indy car scene. Bennett admitted that he had to work enormously hard to achieve an aerodynamic package comparable to the rival cars from Bicester. The pan shot of Andretti at speed emphasizes the tiny rear wings used at Indianapolis.*

mental design brief for the new machine. For the first time, carbon fibre composite materials were employed, the upper section of the monocoque being made from this material, although the lower portion and floorpan were still crafted from aluminium honeycomb.

March also introduced pull-rod suspension, but a spate of rear pull-rod failures saw some teams revert to a rear rocker arm set-up in the interests of added stability in fast corners. By the end of 1983, March had virtually cornered the Indy car customer market, and with no fewer than 30 of the Bicester company's cars in the field at Indianapolis, the new Nigel Bennett-designed Lola T800 was the only rival that looked like standing in the way of a March avalanche, particularly as their ranks were swollen by the Penske team's decision to buy 84Cs and shelve their own troublesome PC12s for 1984.

'We had to spend a tremendous amount of time and effort in the wind tunnel just to match what March had been doing', recalls Bennett, 'because it took so much effort to register just a small improvement.' That said, the Lola really did have genuine winning potential as evidenced by the fact that Mario Andretti turned in a stupendous 212.414 mph (341.774 km/h) lap during unofficial practice prior to the first qualifying weekend. Generally speaking, the long-tailed T800 was marginally faster than the 84Cs on top speed, Andretti tripping the beam at a 222 mph (357 km/h) best on his quickest lap.

In 1985, Lola stepped up the pace of its challenge with

Rick Mears at the wheel of the Penske team's March 85C, a 'customer car' in little more than the sense that it was not originally built by the Reading, Pennsylvania-based team. The Penske outfit always incorporated their own special modifications into the machines which made their March Indy cars less run of the mill than most.

Mario Andretti's T900 coming within a few seconds of winning the Indy 500, finishing just over 3 seconds behind Danny Sullivan's Penske March 85C after a heart-stopping race in which the former Tyrrell driver had survived a spin right in front of the track racing veteran. More worryingly, the lap speed spiral was continuing, and throughout that summer a lengthy and sometimes acrimonious debate continued to rage as to how the regulations should be further modified.

CART eventually announced a revised package of regulations which was supposed to last the Indy car world through from the start of 1986 to 1989. Despite an extensive lobbying campaign by some of the drivers who were anxious to see the Indy car fraternity following the example set by their F1 cousins and adopt flat-bottomed chassis regulations, CART opted to stick with ground-effect regulations.

However, the under-body venturi area was to be reduced by 30 per cent, with the added stipulation that the aerodynamic tunnels would not be permitted to extend beyond the centre line of the rear wheels. There was also to be a requirement for a 'visibility box' between the bodywork and the rear wing, enabling the drivers to see more through their rear-view mirrors.

Side pods had rightly come to be regarded as an integral element within the driver protection package, the minimum and maximum widths across the pods being set at 50 in and 63 in respectively. For the super-speedways there was to be a rear wing width maximum of 56 in (a reduction of 7 in) and engine manifold pressure was left at 48 in. Most significantly, it was announced that the lower half of Indy car monocoques must be manufactured from aluminium honeycomb – not carbon fibre composites – in a style pursued by the latest two March designs.

From a structural viewpoint, one of the most interesting machines never to race was the Gerard Ducarouge-designed Lotus 96, a development of the F1 Renault turbo-engined 95T, which was commissioned by English-born US businessman Roy Winkelmann. In the late 1960s, Winkelmann had owned a private F2 team, operated by driver/manager Alan Rees, which gained an outstanding reputation for high standards of competitiveness and turn-out, and for whom the dynamic Austrian Jochen Rindt won many races.

Winkelmann Racing closed its doors in 1969, Roy thereafter building up a highly successful empire which included security, counter-intelligence and night club businesses. In 1984, he saw his opportunity to return to racing with a CART programme, approaching Lotus to design and build the chassis and Cosworth, rather than an outside engine preparation specialist, to prepare the DFX engines for him.

The original plan envisaged a three-year programme, and Ducarouge paid a visit to the Meadowlands road circuit in New Jersey, just across the river from New York, before producing a 95T-derived chassis concept complying with the Indy car constructional requirements.

The basic carbon/Kevlar composite sandwich pre-preg skins were used much as before, but to imbue the structure with improved anti-sheer strength, the original void-filling Nomex paper foil honeycomb was now replaced with lightweight aluminium honeycomb. Such material needed to be handled more delicately, but was much stronger when bent. In this respect, the 96 marked a new beginning, as the subsequent F1 tubs would also employ aluminium honeycomb filling between their carbon/Kevlar sandwich skins, in place of the Nomex which had been used in all Lotus F1 CFC chassis up to that point.

Unfortunately, lack of sponsorship and driver interest would kill the Winkelmann project before the car ever turned a wheel in anger, and changes in the chassis construction regulations for 1986 rendered its chassis construction illegal in any case.

Although Penske made the switch to full carbon fibre composite chassis construction in time for the 1991 season, Indy car's prime customer supplier would continue to build its chassis with a honeycomb bath tub in which the driver sits. Even into 1993, Lola Chief Designer Bruce Ash-

Above and right *Process of evolution. Two years separate these two shots of Penske cars winning the Indianapolis 500. Emerson Fittipaldi (20) used this Patrick Racing PC17 to win the 1989 event while Rick Mears (3) drove the works PC20 to win the 1991 edition of the Memorial Day Classic. Only very close examination reveals any significant detail changes to the aerodynamic or overall configuration of the cars. Both are in low downforce configuration with aerodynamic wheel centre discs for further reduced drag. Common to both is the amount of protection afforded to their drivers by the high-sided cockpit regulations demanded in this category.*

more continued to harbour doubts about the strength of composite materials in high-speed, secondary impacts. This viewpoint restrained him from committing the company to an all carbon fibre composite Indy car chassis. He was quite firm:

'We've still got a long way to go before we move in that direction. For one thing, it will probably take Penske and the other car builders five or six years to see the same number of crashes that we see, just because we've got so many more cars out there. Those teams are also able to run very tight operations and they run very good, very skilled drivers that are probably less likely to get in the number of accident that we see.

'Nevertheless, you can't protect against having a hub failure or a wheel failure, or something which will cause you to have a major accident. At the moment, we just don't know enough about all-carbon chassis hitting the walls twice at 200 mph.

'You have to bear in mind that in today's world, all-carbon chassis are cheaper to make than honeycomb aluminium chassis, and I don't think you should put driver safety at risk just because it's cheaper to make. We've found a way to mould aluminium to shape like a carbon piece, but it is expensive.

'We build all-carbon chassis for the other formulae we race in, but they don't hit walls at 200 mph. One of our cars crashed last year [1991] at Indy, went into the wall head-on, and then slid down the track. We could see from that that an all-carbon nose is the way to go as it goes through a nice progressive deformation. But further back, when you've got a wheel throwing itself into the side of the cockpit, it cracked the carbon. The crack went down to the aluminium sheet, which then stretched. At that part of the car you need a folding, bending action to absorb the impact, and I think this

is where our cars are better, safer than the others.
'It's very easy to build these cars down to the weight minimum, which is kept at a sensible number. That's the way it should be, so you're not skimping on safety items. The rules have been worded very carefully on carbon structures in order to get the weight of the carbon pieces to be the same as an aluminium chassis. So I don't see why we need to build an all-carbon chassis. Until we're sure it's safe, we won't do it.'

Bruce Ashmore's immediate predecessor as Lola Indy car designer (1984-7) was former Team Lotus F1 Chief Engineer Nigel Bennett, now in charge of design and development at the Penske's team's British base at Poole, Dorset. Reflecting on the manner in which Indy car technology has accelerated over the past decade, Bennett remembers:

'The first Indy car design I did was the Theodore chassis we supplied to Bignotti in 1983. Then we did, I think, three or four days at the MIRA wind tunnel with a very crude one-third scale model, and I think Lola were doing some similar tunnel work at around that time, not an awful lot. But I guess it was from 1984 onwards that things really accelerated from this viewpoint of Indy car design.'

In an effort to control the degree of downforce generated by Indy cars, the underbody regulations have been progressively revised over the past decade to raise the skirt heights. Initially they had to be level with the main floor plan of the underbody. Then the requirement was changed for them to be 1 in above that level. Now it is 2 in. Bennett again:

'They are still full ground-effect designs, but with height restrictions and dimensional limits on the size of the exits from the aerodynamics tunnels, Indy car regulations continue to be very strict indeed. The technical rules form a wad of paper which, I would say, is about four times as thick as the F1 regulations. Everything is tied down in a much more specific fashion than in Grand Prix racing.'

With the modern Indy car required to perform to equal effect on street circuits and super-speedways, the cars are operating across a wide range of aerodynamic and mechanical specifications.

'It's like any other form of racing in the sense that you are always looking for the best compromise between downforce and low drag: and low drag becomes a bigger parameter the higher the speed of the track concerned.

'So at Indy one is looking for a very efficient package where you need quite a lot of downforce, whereas on a slow road course, one's just looking for the maximum downforce the rules will allow. There again, we are much more limited in terms of rear wing dimensions than in F1 – an Indy car rear wing must fit in a box 9 in deep and 20 in long – which means it has quite a shallow chord and is relatively long compared with F1 wings – and we can't run secondary wings like those permitted on Grand Prix cars. As a result, an Indy car produces considerably less downforce from its conventional wings, but considerably more from the underside of the car.'

However, in Bennett's view, the main bonus with an Indy car compared with a flat-bottomed Grand Prix machine, is that they are nowhere near as sensitive to pitch and heave in their aerodynamic attitude. Consequently they can run softer springs and thus have much more suspension travel.

In that respect, Ayrton Senna's test outing in a Penske PC20 at Arizona's Firebird Raceway during December 1992 proved extremely revealing. Senna reported to the Penske team that, in medium-quick, third gear corners, 'The car just invited you to go faster and faster.'

The Brazilian added, 'If that had been the McLaren (MP4/7A) it pitches if you lift off, squats when you get back on the power. You are correcting it all the time.'

Interestingly, another insight on F1 versus Indy cars on the handling front came from Michael Andretti after his first test in the MP4/7A. Suffice to say, the 1991 Indy car champion felt it was quite a handful and came away from

Runner-up in the 1991 Indy 500 but overall winner of that year's PPG Indy Car World Series was 28-year-old Michael Andretti in the Lola T91/00 entered by Newman/Haas Racing.

the test with his respect for Senna's ability significantly enhanced. Bennett:

'We use between 700 and 2,000 lb spring rates, dependent obviously on the wheel rates, but whereas on a Formula 1 car you really have to control the attitude very carefully from an aerodynamic point of view, on an Indy car this is far less critical, so you can go quite soft for wheel adhesion and grip with the result that they are much nicer to drive than Grand Prix cars.'

Lola boss Eric Broadley echoes the diversity of technical challenge offered by Indy cars:

'At Indianapolis, aerodynamics are really critical. We are looking for tiny drag reductions which you are not going to

British technology wins again as the Bicester-built Galmer-Chevy of Al Unser Jr wins the 1992 Indy 500 by 0.043 sec from Scott Goodyear's Lola T92/00.

do in F1. At Indy, you are running at getting on for 230 mph on the straight, and if you take off a wee bit of drag, you're suddenly doing 232...

'Handling is very important as well. The actual cornering characteristic is crucial because the amount of speed you scrub off and the stability of the car in a corner are vital. It's not just an easy flat-out blind. You run with very little wing really slicked off, so the car isn't really planted in the corners.'

The cars run relatively stiff springs at Indy, softer for the road circuits. They also run different wings and aerodynamics at the super-speedways, as well as slightly different top bodywork. Engine cooling is not a problem at 200 mph plus, even running with the smallest radiator ducts. As far as where to position the vents to expel hot air from the radiators are concerned, Lola favoured side ducting in 1992, but Nigel Bennett felt this could never be made to work as well as top ducting:

'There is quite a complicated aerodynamic regime around the rear wheels, and because of the under-body tunnels, a tight F1-style Coke bottle rear end, with air ducted over the top of the diffuser panel, can't be used in the same way. What works on a Formula 1 car doesn't work on an Indy car – and vice versa.'

It also has to be emphasized that Indy cars run different suspension components for the speedways – the banked turns requiring negative camber on one side, positive camber on the other – while the cars also used a locked differential, and the weight distribution is quite significantly altered. Tyre stagger is used to compensate for the lack of differential action.

This latter aspect causes no problems because the current Indy car weight limit is a generous 1,550 lb – too generous in Nigel Bennett's view, as the 1992 Penske PC20 had to carry 90 lb ballast at Indy, 40 lb on the road circuits. He explains:

'Aside from the danger aspect stemming from the difficulty in securing that ballast, there is also the matter of the additional energy which has to be dissipated in the event of an accident. I believe there is a very strong case for reducing it. We are also looking at ways of reducing speeds at Indianapolis quite considerably by 1995, from around the 225 mph in corners to around 200 mph which, in my view, is quite fast enough for anybody!'

Such a strategy presents secondary problems, of course. To begin with the Indy car designers appreciate that they would have to remove a great deal of downforce in order to produce that sort of speed reduction, and even if it was achieved there is no guarantee that such changes would make the cars safer in the long run. 'You may end up having your accident at a speed 20 mph lower, but, then again, you may have more accidents', remarks Bennett wryly.

Penske's switch to all carbon fibre composite chassis in 1990 tends to suggest that Bennett is more audacious in his technical thinking, but in fact his observations on the state of Indy car chassis technology mirror the same guarded comments as those of Lola's Bruce Ashmore. He acknowledges that Indy car designers were initially wary of using carbon fibre composite construction.

'It is a very brittle material, and although it is very strong, when it fails, it tends to fail calamitously. An aluminium structure tends to crumple in a nicely progressive way. So we don't use carbon fibre alone in our chassis; we use it in conjunction with a lot of Kevlar which tends to behave in a more metallic sort of way.

'I note that Lola still believes in keeping an aluminium honeycomb area around the cockpit wall on their 1993 cars, but with the 1993 regulations requiring thicker chassis wall specifications, I think we can still manage to make a safe structure using our techniques. In terms of extra material, the 1993 chassis is 20 lb heavier than the PC20.'

In order to allay any lingering doubts, Bennett recalls that Penske carried out a lot of tests on carbon composite and aluminium boxes, and by experimenting with the lay-up techniques and resin systems, arrived at a design which would fail relatively progressively. However, he freely admits, 'Ultimately, I think I was probably swayed by the fact that a moulded carbon fibre monocoque makes it so much easier to get the shape you want, rather than in a folded aluminium structure.'

Of course, Bennett is realistic enough to know that Indianapolis and the other super-speedways are something of a special case from the standpoint of car security.

'As far as impacts are concerned, Rick Mears's accident during last year's Indy 500 saw the front of the chassis break off, which isn't desirable', he says with masterly understatement.

'You can never be sure in those sort of accidents that the car

A 1992 Penske PC22 nearing completion at the team's Poole, UK, manufacturing base. The huge cast iron ventilated disc brakes and extra structure ahead of the cockpit pedal line, along with the more sturdy push-rods and wishbones, mark the contemporary Indy car apart from its Formula 1 cousins.

is going to be safe enough to protect the driver. These accidents just don't relate to any sort of accident you are likely to experience in F1 because a guard rail is going to move and absorb a lot of energy. I mean, if the wall (at Indianapolis) moved one inch it would make a huge

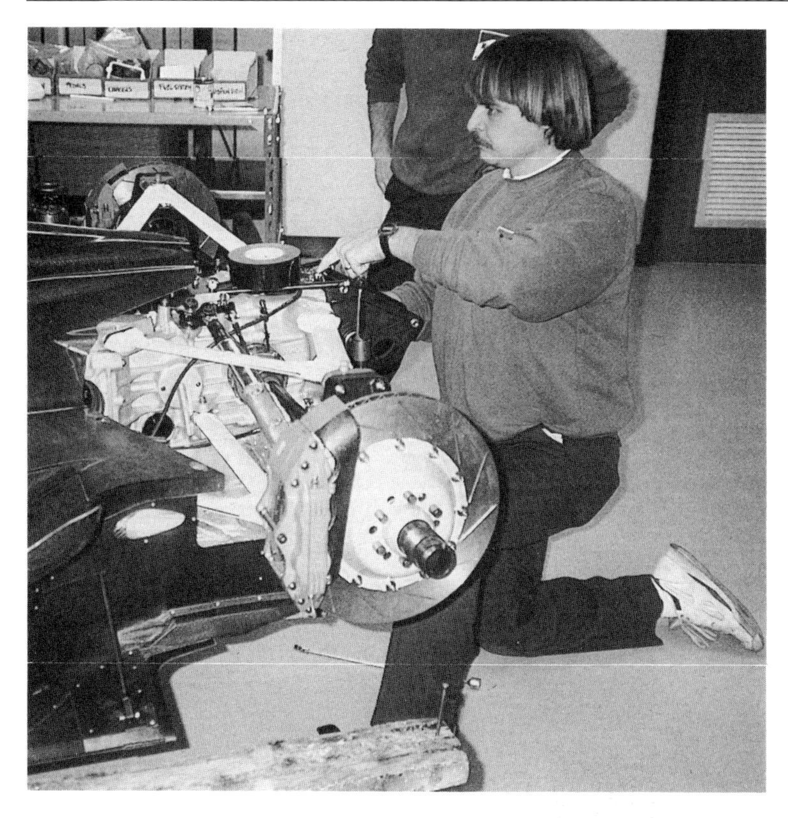

Rear body upper aerodynamics on the Penske PC22 involves careful ducting of air round the rear wheels in conjunction with ground-effect tunnels which are modest in size compared with those on early 1980s F1 cars.

difference, but, of course, it doesn't move at all.

'The saving grace is that the cars are running so close to the wall for most of the time that they don't hit it head-on in the event of an accident. Even Nelson Piquet's accident in practice for the 1992 race wasn't head-on by any stretch of the

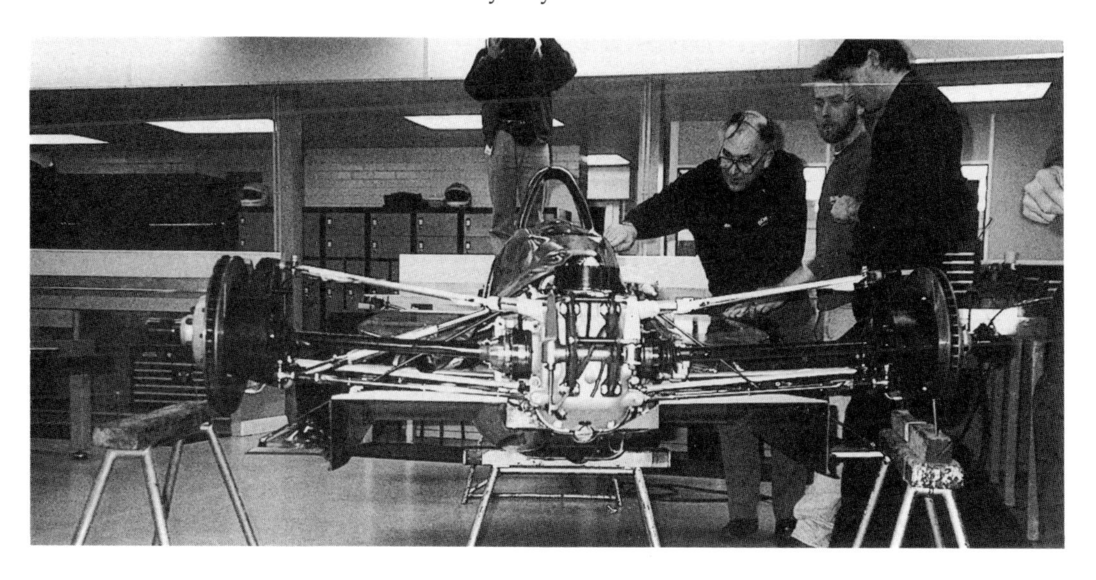

imagination. The forward velocity was probably no more than 70 mph. But that's 70 mph into a solid object. Well, it would kill you if you were in any sort of production car – and I don't think you would stand an earthly in an F1 machine either...'

Indy car impact testing is carried out at MIRA with the same scrupulous attention to detail as is framed within FISA's regulations for F1 cars. Mindful of the fact that the classic Indianapolis-style accident invariably involves a secondary impact, all Indy car nose sections have to be checked with a preliminary impact of 58.5 kJ and a secondary impact of 29.2 kJ with a mass of 900 kg at 10 m/sec. As a comparison, an F1 chassis has to be subjected to a single 47 kJ frontal impact, but the whole of the monocoque now has to be tested, but the Indy car constructors only have to submit the nose sections. Bennett says:

'We also have cockpit rim breakage and cockpit penetration tests, and while we don't have a roll-hoop test for Indy cars, that area of the car is very rigidly specified in terms of

This page and opposite *Penske designer Nigel Bennett working with a half-scale model of the PC22 in the University of Southampton wind tunnel facility prior to the 1993 Indy car season. The model is suspended on a strut which contains sensors to transmit readings to computer screens in the control room, and has a control mechanism which can be pre-programmed to run a series of tests at differing ride heights without the need for time-consuming manual adjustment of the model. The University of Southampton tunnel is a state of the art facility with the crucial rolling road facility long regarded as a key factor in obtaining accurate readings.*

materials that must be used, so there is no real chance of skimping in that area.'

Bennett's Penske PC20 was the first of the team's Indy cars to be equipped with a transverse gearbox ahead of the rear axle line, all previous cars being fitted with conventional, outboard longitudinal set-ups. This enabled the designer to make a worthwhile aerodynamic gain while at the same time improving the car's overall weight distribution.

The Penske transmission development team operates under the control of Geoff Ferris, the man who designed the original Cosworth DFV-engined Penske PC1 Grand Prix car back in 1974 and who has remained with the company ever since. Penske design and commission the manufacture of their own cast magnesium transmission casing, and use purpose-made gears from either XTrac or Emco of Milwaukee. These are rather more sturdy than their contemporary F1 equivalents, reflecting the huge amount of torque generated by the lightly turbocharged Ilmor V8 engines. A sequential motor cycle-style fore/aft gearchange is used, controlling six forward ratios on the ovals, five speeds for the road and street circuits. Bennett adds:

'We also experimented with a traction control system, both retarding the ignition and cutting out cylinders, but like the CVT fully automatic transmission system we developed, regulations were formulated to ban them. As far as the gearbox is concerned, however, a Grand Prix team is currently [January 1993] looking at the system.'

CHAPTER 10

INDY CAR ENGINE DEVELOPMENT

The arrival of the Cosworth DFX V8 in the mid-1970s effectively opened the modern era of Indy car engine technology, sounding the death knell for the four-cylinder Offenhauser shortly after this long-lived museum piece had successfully got the upper hand over the production-based Ford V8s.

In the late 1960s, Louis Meyer and Dale Drake worked hard at trimming weight from the Offenhauser engine at a time when they were losing customers to the Ford V8 ranks. However, once Ford had established its engine as a winner, after Jim Clark's dominant performance in the '65 Indy 500 with the Lotus 38, the Detroit car maker contracted Louis Meyer to distribute and service

Ford stormed the bastions of Indianapolis in 1963 with a push-rod version of their Fairlane V8 sedan engine re-engineered specifically for racing purposes. By 1965 the Ford V8, as illustrated here, had been metamorphosed from push-rod to overhead-camshaft configuration. With one of these units, Jim Clark's Lotus won the 1965 Indy 500.

After Dale Drake's death in 1972, his family completed the redesign of the veteran four-cylinder Offy engine, an example of which is seen here, which had been started by Leo Goossen and 1930s technical innovator Art Sparks. This update to the supercharged Drake-Offy programme was bankrolled by car owner Pat Patrick whose crew chief George Bignotti felt this was the way to go. The development programme finally ended in 1979, the veteran power unit being eclipsed by the more fuel-efficient Cosworth DFX.

the unit for the customer market.

Obviously, this brought an end to the Meyer/Drake partnership. Louis Meyer set up his new business enterprise at Indianapolis, leaving Dale Drake in Los Angeles to concentrate on the classic four-cylinder engine, which was now at least 75 bhp and 2,000 rpm behind the Ford opposition. Dick Jones of the Champion Spark Plug company then took a hand in promoting a Roots supercharged version of the engine, but it was an eventual turbocharged version of the Offenhauser which enabled the old engine to reassert itself in the late 1960s.

The Offenhauser engine's near monopoly on the Indy car winner's circle would last through until the 1976 season, when Al Unser's Parnelli introduced an F1-derived Cosworth V8 to the victory rostrum with victory in the Pocono 500. The following year the McLaren M24-DFXs arrived on the scene, and it was effectively all over for the Offenhauser.

Although Cosworth had built up a small batch of short-stroke DFV engines the previous year for Roger Penske, he had second thoughts about the project, and it was not

The Cosworth DFX brought F1-style serviceability and mechanical dependability to the Indy car scene, hitherto dominated by the increasingly fragile, vintage Offy four-cylinders.

until the Vel's Parnelli team expressed an interest that serious collaboration with Cosworth on the DFX project started during the winter of 1974/75.

With the reliability of the Cosworth DFV so firmly established, and the Offy nearing the end of its life, it was perhaps inevitable that an Indy car team would seriously explore the potential offered by the British engine manufacturer. As Al Unser recalled, 'We were having a terrible time with those Offies. We used to puke' em while we were warming them up; the water jackets were the worst part – they used to crack and shift all the time.'

According to author John Blunsden, the turning point came when all but three of a batch of Offenhauser blocks the team had bought failed to pass a specially arranged pressure test. Team chief Vel Miletich, who was Parnelli Jones's partner, was said to be so incensed by this, allied to the team's inability to gain what he regarded to be satisfactory compensation, that he badgered Jones and other team personnel into pursuing the idea of a turbocharged short-stroke DFV.

They decided the project was worth investigating, so

two key engineers, the young engine builder Larry Slutter and experienced machinist Chick Hiroshima, were hived off from other duties and spent the best part of a year developing the short-stroke turbo DFV derivative to run on the regulation methanol fuel.

By retaining the DFV engine's bore of 85.6 mm and reducing the stroke from 64.88 mm to 57.3 mm it was possible to lower the V8's displacement from 2,993 cc to 2,645 cc, which brought it within the Indy car regulation range of 2.65 litres. Having surmounted initial frustrating piston problems, oil scavenging difficulties and the need for the converted V8 to accept a much increased water flow rate for cooling purposes, the first engine was ready to run in early 1975. Al Unser was highly impressed from the outset as the early units were pushing out in the region of 750 bhp. As he recalls:

'At that time we had 80 in of boost, and the Cosworth had so much power, and was so much smoother, that there was just no comparison. The old Offy shook, vibrated and had a very abrupt feeling coming out of the corners, so if you didn't have the chassis set up exactly right, the car just broke loose...'

The first such engine appeared in the Parnelli VPJ6 chassis in practice prior to the '75 Indy 500, and Al Unser qualified a VPJ6B-DFX on pole position for the engine's first race outing at Phoenix, at the start of the following year. It finished fifth on its maiden outing – a race won, ironically, by Al's elder brother Bobby in an Eagle-Offy – but the Parnelli went on to score commanding wins in the Pocono 500, Milwaukee 200 and end-of-season Phoenix 150.

By the start of 1977, Cosworth were becoming heavily involved in the programme, realizing just what commercial potential the engine offered in this new market. As Keith Duckworth told author Graham Robson:

'I was in one of my "Well, we don't want to expand" periods. We only went into Indy racing because I thought that our F1 business might be dying. Then, instead of dying, our F1 business stood up well, the DFX engine took off like a rocket, and that caused us to expand rapidly once again.'

Duckworth's concern about the DFV's F1 future had been triggered by the arrival of Niki Lauda and the Ferrari flat 12 on to the Grand Prix Championship stage, but although this combination would be the one to beat between 1975 and '77, Maranello's F1 pre-eminence was not destined to last.

Suddenly, the Offenhauser was facing extinction. Of the 14 races making up that year's Indy car schedule, Cosworth DFX engines would power the winner on eight occasions, six of which were achieved with the V8 installed in the new McLaren M24, two with the Parnelli. In 1978 Penske and Lola joined in as DFX users with their own Indy cars, and the Cosworth V8 was effectively left in command.

Interestingly, the DFX was always badged a 'Cosworth' in the United States rather than a 'Ford'. As Ford's competitions boss Walter Hayes explained at the time:

'I actually suggested to Keith that it should be called a Cosworth because there were people in the USA who wanted to run it, but who had contracted arrangements which precluded them for going with a Ford engine – for example, it would be very difficult for a Chevrolet dealer to do so.'

The argument about boost pressure levels which formed one of the cornerstones of the great CART/USAC controversy in 1978/79 briefly saw a threat to the DFX domination from Chevrolet's all-aluminium 5.8 litre stock-block V8. But reductions in boost pressure from 80 in in 1978 to 50 in in 1979 and then 48 in for 1980 allowed the DFX to develop a reputation for admirable reliability. The old four-cylinder Offy was permitted to stay in play with a 60 in boost level, but even afforded that luxury, the Indy old-timer had breathing problems which prevented it from developing sufficient power to handle the DFXs, even with the V8s running at their much lower boost level.

As one of the relatively few drivers to have experience with the Cosworth DFV F1 engine and DFX Indy car units, Mario Andretti had some shrewd observations to make:

'Compared with the DFV, which gave out a relatively modest output but is very responsive, and has a very good power curve, the DFX felt a bit of a beast.

'Trying to use full power and use it effectively was a real struggle when there's no way you can drive smoothly because there's no progression. As far as what's at the end of your foot is concerned, well, it's a bomb at 8,000 rpm and it's a bomb at 11,000 rpm. The way the extra power comes in just knocks you out of shape when you're road racing. In the rain, man, it was a terror.

'When they first cut back the boost, the top end of the DFX came down quite a bit. But all that was quickly made up with cams and compression ratios, and by the early 1980s it had the kind of horsepower that made it a really stout engine. You could really feel the horsepower, whereas when we first started running 48 in of boost, we were down to about 650 bhp, and that was a little too docile.

'Of course, both the DFV and DFX were real vibrators because of the 180-degree cranks, and as far as smoothness and progression were concerned, they were nothing like a 12-cylinder Alfa or Ferrari, although they did give a very efficient feel. Yet on a super-speedway like Indianapolis or Michigan, the DFX operated fairly effortlessly at 11,200 rpm. It must be that the harmonics are just right at this range, because it feels like it could go on forever, and, of course, a lot of times it did just that.'

In 1983, two ex-Cosworth employees did to the DFX what the F1-derived V8 had done to the Offenhauser. By 1989, if you wanted to make the big time in Indy car racing, it was best to choose a Chevrolet, otherwise known as the Ilmor V8.

Mario Illien and Paul Morgan had been employees of Keith Duckworth's august company, each working on a different project. Illien had a major hand in the Sierra Cosworth road car programme, as well as producing the DFY variant of the company's proven 3 litre DFV Grand Prix engine. Morgan, meanwhile, had been responsible for looking after the company's Indy car fortunes as chief engineer on the DFX project in the early 1980s.

At the end of 1983, the two men put their heads together and struck out on their own, establishing Ilmor Engineering on an estate at Brixworth, near Northampton. Together, they decided to take on their former employers. Illien recalls thoughtfully:

'When we left Cosworth we considered carefully which racing category to tackle, and we concluded that tackling Cosworth's monopoly in Indy car racing would make most sense. So we called Roger Penske...'

Penske, the patrician, multi-millionaire driving force behind Indy car's most consistently successful team, listened with interest as Illien outlined the new company's plans and was impressed with the Swiss-born engineer's belief that he could improve on the Cosworth DFX. Illien and Morgan, in turn, were struck by Penske's decisiveness in tackling the project.

'All he wanted to know was how much it would cost and how long it would take to complete the first engine', says Illien. With typical commercial audacity, Penske agreed to underwrite the project, convinced that he could, in turn, persuade a major US corporation to back it. When he successfully romanced General Motors into the project, most observers were simply amazed. Here were plans being laid for a brand new Indy car V8 with 'Chevrolet' emblazoned on its cam covers.

Illien and Morgan began work on the project in November 1983, and by the following March the first drawings for the Ilmor 265A V8 were coming off the drawing boards. Ilmor saw the first castings in September 1984, and on 16 May 1985 the first completed engine burst into life on the test bed.

On 10 August that same year, the test track at Bruntingthorpe echoed to the raucous crackle of a racing engine. Installed in a Penske March chassis, the team's number one driver Rick Mears had his first taste of Ilmor power. Eight months later, in April 1986, Al Unser would give the new V8 its race debut in the new Penske PC15 at Phoenix.

As things transpired, installing a brand new engine in a brand new chassis was not going to prove a particularly productive affair. Although the PC15 would eventually be massaged into half-way respectable form, Al Unser branded its handling 'lurid' on the super-speedways, and for most of the time it took something of a back seat to the team's fine-handling March 86Cs.

The Honda V8 of 1986 wasn't really a Honda at all, despite the fact that it carried 'Brabham Honda' identification on its cam covers. In fact, it was a straightforward Judd V8; in effect a de-stroked 2.65 litre turbo version of the British engine specialist's Formula 3000 unit. The unit was installed in one of Rick Galless's Lolas for Geoff Brabham, appropriately enough, to drive.

The engine was also bedevilled by torsional vibrations which dogged the engine's early development, and, by Illien's admission, it was not until around October 1987 'that I felt we'd really got a handle on this particular problem'. He also made the very generous point that he didn't think running the Ilmor engine in the PC15 chassis helped Penske designer Alan Jenkins a great deal, 'and I know it didn't help us'. At least when the new V8 was installed in the back of a trusty March 86C the team had a chassis performance base line by means of which they could cross-reference the engine's performance against the Cosworth DFX.

Even when running customer March chassis, of course, the Penske team would add their own bespoke modifications to the car, a habit which had been developed when they bought three 85C chassis at the start of the '85 season. It wasn't long before the Poole-prepared cars appeared with revised bodywork, their own wings, beefed-up rear rocker arms and headrest fairings to reduce buffeting around their drivers' heads.

The Ilmor engine scored its first victory in Mario Andretti's Lola at Long Beach in April 1987, and between 1988 and the end of 1991 powered the winner in all but three of the races held on the Indy car schedule. That year (1987) would also see Penske win the Indy 500 but not with the

The Cosworth DFS was an uprated derivative of the original DFX concept introduced in 1987 to combat the increasing challenge from the Ilmor V8s.

Ilmor engine, nor the new Penske PC16 for which it was intended, this car also proving to be a big disaster. Penske hauled out his trusty March 86Cs, bought the previous year, and Al Unser used a Cosworth DFX-powered example to take his fourth career win at the Brickyard.

Competition for the Ilmor-built V8 had come from Porsche, Alfa Romeo, Judd and the by now outdated Cosworth DFX and DFS engines during that period, each of which managed to score a win apiece during 1988 and '89, but the Chevy clearly had the edge on power. Porsche then withdrew from Indy car racing at the end of 1990, followed by Alfa Romeo a year later, and only a single team existed on a diet of Judds in 1991.

Cosworth, however, was not standing still. Between 1988 and '91 it worked flat out on a new V8 engine, code-named the XB, and from the start of 1992 this compact new V8 made its debut powering the Newman/Haas Lolas of Michael and Mario Andretti, plus the cars for Eddie Cheever and the 1990 Indy 500 winner Arie Luyendik.

Design work on the Ford-Cosworth XB began in June 1989, and the engine first ran on the test bed at the end of July the following year. On 23 September 1991, Mario Andretti tested it for the first time in one of the Newman/Haas Lolas, and was absolutely delighted with its potential.

As well as being smaller and around 60 lb (27 kg) lighter than the Chevy V8, the Ford-Cosworth XB produced more power from the outset. Michael Andretti paid it the ultimate compliment when he said, 'It feels just like a Chevrolet, only with more power.' The XB was also able to run safely to 13,000 rpm, giving it a 500 rpm edge over the Ilmor product, and showed from the start that it was slightly more fuel efficient.

In charge of the XB programme from the outset was Cosworth's racing manager Dick Scammell, now a director of the company. He was responsible for overseeing the project, together with engineers Steve Miller and Malcolm Tyrrell.

As Miller explained to respected US journalist Gordon Kirby in *Racer* magazine:

'We knew from other projects how to reduce engine friction, so initially we intended to design an engine that was going to be smaller and lighter, and would produce more power simply from reducing that friction.

'From the development of our 1.5 litre F1 turbo in 1985/86, and from our naturally aspirated V8, we knew we could make the engine more compact, and work we did, during both the design stage and the development programme, led to the engine breathing better and producing more power.

'Early in the design stage we went around and talked to all the car constructors. As a result, we found the general desire was for a slim engine to correspond with the roll hoop and the driver's head. Getting the height of the plenum chamber down wasn't a big consideration. The car builders preferred rear bodywork which was tall and slim rather than something which was low and flat. As a result, the requirements for a slim plenum chamber and a slim sump dictated the angle of the cylinder bank.

'As far as economy is concerned, most of the improvement was achieved in the cam timing and injection area. Apart

The Penske team pioneered the use of the new Ilmor 265A V8, manufactured by the new company formed by ex-Cosworth engineers Paul Morgan and Mario Illien who struck out on their own at the end of 1983. Penske persuaded General Motors to adopt the engine which has carried Chevrolet identification from the outset.

from the reduced friction, there's very little in the fundamentals of the engine that we could address to improve. But we were able to produce a better piston design and to make them out of better alloys than are commercially available.'

Chevrolet, meanwhile, didn't just stand still and allow itself to be overwhelmed. The decision was taken for Ilmor to instigate a crash programme to revise the V8 design into what would become dubbed the type 265B.

Mario Illien played his cards very close to his chest in talking about the revised V8. 'It's just a smaller package', he commented shortly after the engine was run for the first time on the Ilmor dyno in December 1991. 'The concept is very similar. It's just packaged a little neater and smaller. It uses the same pumps, but the rest is all new. It has new block, heads, connecting rods, piston, cams and drivetrain.'

The Ford-Cosworth XB and Chevrolet-Ilmor 265B engines would both make their Indy car debuts at the start of the 1992 season. Ironically, both looked highly promising, but both were plagued with mechanical unreliability. In the event, Bobby Rahal used a Lola powered by the earlier Ilmor 265A engine to win the PPG Indy Car World Series, and a similar unit propelled Al Unser Jr's Galmer to victory in the Indy 500.

The tradition for encouraging stock-block engines at Indianapolis has always been well entrenched, encouraging Buick to pursue a nine-year development pro-

The Ilmor 265A (right) pictured with its successor, the 265B which first ran on the dyno in December 1991. Outwardly similar, the new engine featured new cylinder heads, block, connecting rods, pistons, camshafts and drive-train.

Cosworth's latest tool in the battle against the Ilmor opposition, the slim, compact XB V8 made its debut on the Indy car scene in 1992. With reduced internal friction, better breathing and slightly more power, like the new Ilmor 265B it was beset with mechanical unreliability during its maiden season.

gramme which finally yielded its best result in 1992. At the wheel of a Menard team Lola-Buick, deputizing for the injured Nelson Piquet, Al Unser Sr finished third behind his son and Scott Goodyear's Lola to register the best ever Indy finish for the 3.4 litre, single-cam, push-rod Buick V6, which under the Indy rules had always been allowed an extra 10 in of manifold pressure as compared with the regular 2.65 litre four-cam V8s.

After two years of development in the hands of McLaren Engines in Detroit, the Buick V6 was given its first Indy car airing in the hands of Scott Brayton back in 1984. The following year he qualified second to Pancho Carter's similar March-Buick which had taken pole position at a record 212.583 mph (342.046 km/h).

Qualifying was one thing, but racing the Buick V6 had always proved a very different proposition. Reliability over 500 miles was always a problem for the push-rod unit, churning out around 800 bhp at a lazy 8,000 rpm. Away from the Brickyard, CART rules had always restricted Buick's engine to the same boost pressure as the four-cam V8s, believing that the extra cylinder displacement was sufficient to balance the design differences. However, for 1992, the Buick was allowed to run in all Indy car races with 50 in of manifold pressure, 5 in more than the V8 opposition. Not that it made much difference, mind you, for Unser's Indy success stood unchallenged as the engine's best result.

CHAPTER 11

SPORTS RACING CARS

Porsche sets the Group C scene

World Championship sports car racing found itself locked into a terminal decline by the end of the 1980s, eventually fading from the international calendar at the end of 1992. Many historians regard the seeds of its demise to have been sown as long ago as 1971 when the CSI, which pre-dated FISA as the sport's governing body, initiated a maximum engine capacity of 3 litres, thereby outlawing at a stroke the superb 5 litre Porsche 917s and Ferrari 512s which had been central players in this category throughout the previous four seasons.

The result was a category which played straight into the hands of the French Matra organization, who duly went on to score a hat-trick of Le Mans victories between 1972 and '74, thereafter quitting this category of the sport in favour of a role as engine suppliers to the F1 Ligier team. There followed a period of piecemeal, unstructured evolution for the rest of the 1970s, the CSI gallantly attempting to patch up the mess they had originally created by writing rules which seemed calculated to accommodate any player with a reasonably competitive car. But in 1982 the new Group C regulations were introduced with the intention of creating an easily policed international sports car category.

The permissible dry weight of the cars had to be not less than 800 kg, and although ground-effect tunnels were allowed in much the same way as on Indy cars, aerodynamic side skirts were not permitted to bridge the gap between the outer extremity of the bodywork and the track. Moreover, a flat plate beneath the cockpit, measuring at least 100 cm by 80 cm had to be incorporated into the design as a further means of reducing the downforce.

Any engine was permitted, providing it came from a manufacturer which had cars homologated into Groups A or B, but the most important aspect of the regulations was that the cars were limited to five refuelling stops within a 1,000 km race, or to 25 stops in the Le Mans 24-hour classic, which still remained overwhelmingly, the jewel in the sports car racing crown. Moreover, FISA followed the sample of the Le Mans organizers by requiring that refuelling be fixed at the rate of 50 litres per minute, so that such stops would necessarily last for a couple of minutes or more.

The main player during the initial Group C season was Porsche with its type 956, which, along with the subsequent 962 development, would be the archetypal 'customer car' for much of the next 10 years. Not until Tom Walkinshaw's Jaguar operation came on the scene in 1986 was this status quo significantly disrupted.

The longest lived Group C car of the 1980s was the privateers delight, the Porsche 962. This is the Repsol-backed machine of long-time Swiss Porsche privateer Walter Brun, which came within minutes of winning the Le Mans 24-hour classic as late as 1990.

Porsche's Group C programme was initiated at lightning pace over a period of nine months towards the end of 1981, after the green light for the project was given by incoming chief executive Peter Schutz. The Weissach design team decided to take advantage of the maximum possible overall dimensions permitted by the regulations governing the construction of their new car, the rationale being that the greatest amount of possible floor area

would maximize the available ground-effect possibilities, but there was a trade-off in terms of weight, and no Porsche 956 would ever quite slim down to the 800 kg weight limit.

By the F1 standards of the time, the Porsche 956 was regarded as something of a dinosaur, for while its Grand Prix cousins were increasingly being built around carbon fibre composite chassis, the new Porsche was made from aluminium sheet, bonded and riveted into monocoque form, and stiffened by means of an aluminium roll cage. That said, this structure proved to be immensely durable, as evidenced by the fact that the factory's own engineers thrashed the first prototype for 1,000 km over their pave test track section without any failures.

The Porsche flat six, 2.65 litre turbocharged engine ran at a modest boost level of 1.2 bar in race trim – a far cry from the unlimited boost employed in F1 until restrictions were applied in 1986 – and finished the first season developing about 620 bhp economically and with impressive torque characteristics.

The finished result was clad in a seven-piece bodyshell manufactured from carbon fibre-reinforced Kevlar. There wasn't much Porsche could be told about aerodynamic development, of course, and the 956 displayed impressive CD figures ranging from 0.35 in low-drag Le Mans configuration to between 0.40 to 0.42 in higher downforce trim, a testimony to the amount of work carried out in the VW wind tunnel and the similar Mercedes-Benz facility in Wolfsburg.

A one-fifth scale model of the 956 had been ready for initial aerodynamic assessment in Porsche's own wind tunnel a mere 55 days after the project had been given the go-ahead, but the company had to look outside to the facilities of its fellow German car constructors in order to continue with full-sized model testing.

Since fuel consumption was such a fundamental element within the Group C equation, Porsche became involved in a major programme of collaboration with Bosch, which resulted in the development of the Motronic 1.2 system in time for the works cars to use in 1983, although this luxury was not made available to the marque's private customers until the following year.

The US goes it alone

To the disappointment of the European manufacturers, the USA was not destined to adopt the Group C regulations, and the International Motor Sports Association (IMSA), which presided over the most significant sports car series in the USA, chose to go its own way. This was founded on a belief that the American viewing public would take little account of the fuel economy message which was implicit in the Group C regulations, and would certainly not enjoy contests in which fuel efficiency seemed calculated to slow the pace.

IMSA supremo John Bishop felt that the US sports car racing fraternity, which was dominated by private entrants, needed something different. His proposed new Grand Touring Prototype cars were destined to be very similar in construction to Group C machines, but their performance would be governed by a sliding scale of car weights, which was directly related to the size and configuration of the engine.

As an example, a car powered by a 3 litre racing engine such as the Cosworth DFV would have to weigh in at a minimum of 900 kg, but anybody using the 3.9 litre Cosworth DFL derivation would have to field a heavier car. A Porsche 911 turbo, with two valves per cylinder, was given an equivalency of 3 litres and 1,000 kg minimum weight, while any forced induction racing engines were set at 900 kg for 1.5 litre units and 1,000 kg for 2 litre machines. The Porsche 956 was rendered ineligible because of its 2.65 litre nominal capacity, and in any case this Group C machine did not conform to the IMSA requirement that the pedals should be completely behind the front axle line, a factor which was taken into account when the 962 was developed specifically for its US privateers to use in the IMSA series, starting in 1984.

To conform with the American regulations, the 962 was powered by what was virtually a 935/76 engine of 2.8 litres, with two valves per cylinder and a single KKK turbocharger. Fresh from the factory, the 962 tipped the scales at 840 kg, thus requiring 10 kg of ballast to conform to its minimum stated weight.

The 962 made its debut in the 1984 Daytona 24-hours, handled by the father and son combination of Mario and Michael Andretti, and was quickly followed by the arrival of four similar machines on to the IMSA scene early that year.

The main apparent advantage of IMSA's control of its domestic series is a readiness to adjust the regulations in order to prevent any one machine from running away with all the races. In 1986, the goalposts were moved again with a new sliding scale of weights and engine capacities, unsurprisingly after Porsches won 15 of the 16 races the previous year, but the German marque stayed on top the following year with 12 wins on the IMSA trail, a fact which mainly reflected their pure weight of numbers.

The net result of all this fiddling with the regulations was to make the IMSA series a potentially attractive series to new teams, Nissan and Jaguar joining in to great effect in the mid 1980s, as indeed they would on the Sports Car World Championship stage.

ASPECTS

Carbon-fibre technology

The works Porsche 962Cs, with their pedal boxes positioned behind the front axle line to oblige FISA's decision to align this aspect of its chassis construction rules with IMSA at the start of 1985, made its debut at the Mugello race that year.

However, the Silverstone-based privateer Richard Lloyd, who had commissioned a special chassis for his earlier 956, based around an aluminium honeycomb structure, complete with fully adjustable F1-style suspension, repeated the exercise when he got his hands on a customer 962C.

The honeycomb 956 was praised by all those who drove it, the feeling being that it displayed significantly enhanced torsional stiffness as compared with the standard sheet-aluminium monocoque, although the Lloyd

Many Porsche 962 privateers pursued their own line of chassis and aerodynamic development, one of the most ingenious of such teams being the Silverstone-based Richard Lloyd Racing. This is his special 962C which was entered in the 1989 Silverstone 1,000 km race for Derek Bell and Tiff Needell. Lloyd had earlier commissioned designer Nigel Stroud to build the team's own aluminium honeycomb and later carbon fibre composite chassis on an exclusive basis.

machine was about 20 kg heavier than the standard product.

It was clearly not going to be long before carbon fibre composite materials were applied to the construction of a Group C sports car. Eventually it would be the new TWR Jaguar team which set the pace on this particular front.

The trailer for Jaguar's splendidly successful assault on the sports car racing scene during the second half of the 1980s was signalled by the US-based Group 44 organization which had been presided over by Bob Tullius for many years. In 1981 this enterprising operator, who had for some time been involved in racing Jaguar products in minor league US categories, realized that the IMSA series would provide the catalyst for involving the British manufacturer in the sports car racing big time.

Jaguar takes over

The first of what could be described as the contemporary racing Jaguars was the Lee Dykstra-designed XJR-5, a sheet aluminium semi-monocoque coupe, powered by a 5.3 litre version of the XJS V12 production engine. This made its race debut in the summer of 1982, to be followed in 1983 by a 6 litre V12, developing around 600 bhp, which proved competitive enough to win four IMSA rounds, and the following year Jaguar Chairman John Egan was persuaded to support the idea of bringing a pair of Group 44 XJR-5s to Le Mans.

Neither car finished, but they both showed up extremely well indeed, and British team owner Tom Walkinshaw, who had done great things for Jaguar with an XJS coupe in the European Touring Car Championship, was asked by Egan to evaluate the XJR-5 for his forthcoming World Championship sports car programme.

However, Walkinshaw had his own agenda firmly in mind and was not about to base the most important racing project of his entire career around somebody else's hardware. Although Lee Dykstra had a new XJR-7 IMSA design almost finished on his drawing boards, offering more downforce and lower drag, using a degree of carbon fibre and Kevlar in its chassis construction, the Group 44 effort remained confined to the North American scene, thereafter, closing down the programme at the end of 1987 when Jaguar withdrew its support and lined up wholeheartedly behind Walkinshaw's efforts.

The go-ahead to build the first of the Walkinshaw Group C Jaguars was given by the Coventry company's board in February 1985, although by that time the canny Scot was well advanced with the design concept. In the autumn of the previous year he had commissioned British designer Tony Southgate, a man with considerable F1 experience throughout the previous decade, to initiate work on the first car.

Southgate built the new TWR Jaguar XJR-6 around a chassis constructed entirely in carbon fibre and Kevlar materials, the rear cockpit bulkhead being recessed by 4.3 in (11 cm) in order that the long 60-degree V12 engine could be accommodated significantly further forward to enhance the machine's weight distribution.

This was a beautifully made package which was responsible for raising the sports car racing stakes, and indeed comprised several features – including a central, single fuel cell (from 1986) – and Southgate also drew on his F1 ground-effect knowledge by incorporating very tall aerodynamic tunnels, facilitated by the use of the tall, narrow engine which would help endow the finished product with levels of downforce that the Porsche 962C drivers were certainly not accustomed to.

This venturi effect was maximized by the inclusion of spats over the rear wheels to reduce the escape of air, and while the British-made 6 litre V12 was capable of churning out around 630 bhp, driving the rear wheels through

The Tony Southgate-designed Jaguar XJR5 not only set new standards of Group C aerodynamics on its introduction in 1985, but also featured more extensive use of carbon fibre composite construction than its predecessors and rivals.

March gears housed in a purpose-built TWR casing, it was heavier by 40 kg than the rival Porsches. In an effort to narrow the performance gap further, a 48-valve version of the V12 was produced, but while this topped the 700 bhp mark, it was put on one side for the time being as its added weight and complexity were not considered to be worth the effort.

Despite weighing in at a little over 900 kg, the race debut of the TWR Jaguar XJR-6 was outstanding, Martin Brundle pulling away from the Porsche opposition in the early stages of the Mosport Park endurance race in August 1985. A wheel-bearing failure thwarted his progress after only 13 laps, but Jaguar had signalled that they meant business and were on the Group C scene for the long haul.

After their Canadian debut, the XJR-6s continued to compete in the handful of races through to the end of the season, this restricted campaign yielding a second place best on the Shah Alam circuit at Selangor in Malaysia.

For the 1986 season, TWR expanded the engine's capacity to 7 litres, at which point it developed around 720 bhp, while at the same time the V12 was subject to a rigorous programme of lightening. On the chassis front, the incorporation of the central fuel cell, the removal of the on-board jacking system and more work on lightening the bodywork brought the XJR-6's weight down to around the 860 kg mark. By contemporary Porsche standards, the Jaguar was developing into a very sophisticated racing car indeed, and now generated in the order of 70 per cent more downforce than the German cars.

That season would see Derek Warwick and Eddie Cheever win the Silverstone 1,000 km endurance event, the first Group C success for the TWR squad. Despite failing to finish at Le Mans, the team came close to winning the World Championship for Sports Prototypes, as FISA had chosen to title this major Group C series, and the TWR team went into 1987 feeling particularly confident, since fuel additives, including toluenes, which increased the permissible fuel density to the advantage of their turbocharged rivals, were banned from the start of the new season.

Despite FISA's continued dabbling with regulation changes, a habit the sport's governing body could never quite rid itself of and one which would eventually contribute to the demise of world class sports car racing on a serious basis, Jaguar stamped its mastery on the Championship stage and took the 1987 title. However, its evolutionary XJR-8s again missed out on the jewel in the crown, only managing to get a single survivor home in fifth place at Le Mans.

The superb structural integrity of the XJR-8s central carbon fibre/Kevlar monocoque section was graphically illustrated during the French 24-hour classic when Win Percy ran over a piece of debris on the Mulsanne straight and burst a tyre. The car was completely wrecked in the ensuing high-speed accident, but the cockpit cell did not distort and the driver emerged totally unhurt.

For 1988, having delivered Jaguar the Championship, Walkinshaw expanded the programme to encompass an IMSA programme as well as TWR's defence of the world title. Tony Southgate embarked on a further campaign of development, and the resultant XJR-9 featured a larger 120 litre fuel tank to conform with the regulations for the forthcoming IMSA confrontation, where the cars would run with 6 litre engines as opposed to the 7 litre units retained for the Group C Championship.

Other changes for the US series involved the use of equal diameter wheels front and rear, a quirk which demanded that Southgate redesign the rear suspension, tucking the uprights and dampers well inboard, since the Group C suspension would not fit within a 16 in (41 cm) rim. The Group C cars also ran smaller 17 in (43 cm) diameter rear rims to facilitate a lowering of the rear bodywork in the interests of enhanced aerodynamic performance.

Although the Electramotive Nissan GTP turbo would dominate the IMSA series to Jaguar's detriment, the TWR-operated cars would not only retain the Group C Champi-

The splendid Jaguar XJR19 complete with aerodynamic rear wheel spats seen at Donington Park in 1989 at the height of the TWR team's domination of the Sports Car World Championship scene.

onship, but also won again at Le Mans, Jaguar's first success at the Sarthe circuit since the last D-type victory some 31 years before.

Having been the dominant force in 1988, Jaguar slipped from the high wire in 1989 when the 7 litre XJR-9 machines received their come-uppance at the hands of the 5 litre Mercedes-engined Sauber C9s. The reality was, in its fifth season, the TWR Jaguar V12 engine and chassis package had reached the end of its competitive road.

For some time Walkinshaw's team had been convinced that a turbocharged engine was the way to go, initially for the IMSA series and with a Group C version to follow. However, the abiding rationale behind the Jaguar's sports car racing involvement from the outset was that a production-based engine be used, so in order to secure the blessing of the company, Walkinshaw agreed that the turbo would have a future road car application. As he explained:

'We did some tests on the in-line AJ6 engine just to see what power might be available from a turbo, and this confirmed our belief that forced induction was the way to go. We had some theoretical calculations about what sort of power level might be available, and everyone [in IMSA] was claiming that they were getting a lot less than that. So we just ran it to

The superb 7 litre TWR-developed Jaguar V12 installed in an XJR9 chassis and displaying the way in which the exhaust plumbing was routed to avoid the ground-effect tunnels extending inside the rear wheels.

prove our calculations and then went right ahead with the new engine project.'

The engine, produced by a team headed by TWR's engine supremo Allan Scott, was new from the ground up, although derived from the V6 Austin Rover engine, for which Walkinshaw's organization had acquired the rights some years earlier. In fact, all TWR used that unit for was to carry out preliminary development work on the engine management systems, and the new engine represented a totally separate project, eventually finding its way into the Jaguar XJ220 supercar.

Walkinshaw cited the apparent road car application as the reason behind the new engine having a 90-degree vee angle, rather than what he described as an ideal 60-degree configuration for racing purposes:

'However, the car is so small that we could get both the IMSA and Group C (aerodynamic) tunnels in the car with either vee angle. A sports car is much wider than an F1 machine, for example, so the width of the engine is not really a controlling factor.'

Walkinshaw also went on to explain that the whole XJR-11 package should be technically more effective, from a weight, packaging and aerodynamic standpoint. Referring to the trusty V12, he explained:

'For the last few years we've had a production engine, a really fantastic, but large, tall production engine. Because it's a production engine,it is big and heavy and that has had an influence on the handling of the car. The new engine, being very small and compact, should give us a lot more scope to tune the chassis than with the V12.

'The V12 started off as a good engine, we've worked very well with it and every car we've done we've updated and gone faster. But, inevitably, you end up with compromises; what we were looking for was a small, low and compact *racing* engine, and hopefully that will open up a lot of scope for us on the chassis side.

'The new engine will not give us any more power, or torque, than the V12, but the package which produces that power is much more suited to a racing car.'

From the touchline, the advantages of the new car were immediately obvious from its overall profile. The XJR-11 was much lower than its predecessor with a more

SILK CUT
PUMPS
START

JAGUAR
JAGUAR
3
TWR
JAGUAR
GOODYEAR
GOODYEAR

adequate flow over the rear wing. It was also very light, requiring a considerable amount of ballast to bring it up to the minimum levels required to conform with the IMSA and World Sports Prototype Championship minimums.

For its IMSA applications, the V6 was restricted to a maximum capacity of 3 litres, each turbo having a 38 mm air restrictor under this category's complex equivalency rules. For the WSPC, the ending would run to a full 3.5 litre capacity.

The new car made a sensational debut at Brands Hatch, where Jan Lammers smashed the Group C lap record to qualify the XJR-11 on pole position. However, the fuel consumption of this hastily developed V6 proved excessively high, and although Lammers frequently proved quick again in qualifying, the Jaguars rarely featured in the race results. By the end of the season, while the XJR-11 was acknowledged to be probably the best chassis in the business, it was clear that a lot of work would be needed on the engine management system if it was to be a contender in 1990.

As things transpired, in 1990 the XJR-11 would prove no match for Sauber-Mercedes opposition, but Tom Walkinshaw was quick to identify the problems, and expanded the technical base of his operations, now under the direction of ex-Arrows F1 chief designer Ross Brawn whose main task was to develop the new XJR-14 project to run under the new 3.5 litre naturally aspirated regulations which would come into force from the start of 1991.

Meanwhile, the challenge of developing a fully efficient engine management system for the V64V engine fell to former BMW specialist Gerhard Schumann who made major changes to the overall package, the most obvious being a switch from Zytek to Bosch Motronic engine management system. The team also switched from Dunlop to Goodyear rubber, and was psychologically boosted by the return of Walkinshaw's personal favourite driver choice, Martin Brundle, as team leader.

However, during the second half of the season, the Jaguar effort seemed to falter perceptibly. Ross Brawn, obvious preoccupied with the 1991 programme, no longer appeared at the races, and the team did not send a spare car. The wisdom of being fully prepared for the new 3.5 litre project was unquestionable, but a host of minor technical problems with the XJR-11, including malfunctions with oil pumps, alternators and driveshafts, made it a frustrating time for those involved with the immediacy of racing in the field.

Opposite page *The 3.5 litre V6 turbocharged Jaguar XJR11 was developed for the 1990 Championship and although tremendously fast was beset by acute unreliabilty. The liberal use of carbon fibre composite materials in its cockpit construction can be seen to advantage in the shot 'through the door'. The side gearchange was switched to the centre on the following year's even more compact XJR14.*

The superb Mercedes 5 litre turbo V8 formed the backbone of the Sauber Group C challenge, winning many races as well as enabling Mercedes to learn a great deal about their engine's performance under stress.

For 1991, Ross Brawn took the wraps off his latest design, the 3.5 litre XJR-14, which benefited from Ford's ownership of Jaguar in using the Cosworth-built, but Jaguar-badged, F1 Ford V8 engine as its motive power. TWR opted to use Bosch rather than Ford electronics, but otherwise the engine was essentially the same as its F1 cousins, apart from some concessions to longevity. Brawn explained:

'The XJR-14 is definitely an F1-derived design, as opposed to a traditional sports car package. It is essentially a wide Formula 1 chassis with bodywork that covers it wheels. Externally, it looks like a sports car and conforms with the defined dimensions. But with the bodywork removed, it is a two-seater F1 car.'

Design work on the XJR-14 had in fact started two years earlier with Brawn repeatedly resting the concept in the wind tunnel during the course of its development programme. 'Under-body aerodynamics and wing design remain crucial to successful sports car performance', he asserted, 'given the constraints we have on the upper body.'

There were other major design features which broke with established sports car racing tradition on the XJR-14.

Teo Fabi strapped snugly into the cockpit of the Jaguar XJR-14 in which he won the 1991 Sports Car World Championship.

The TWR-designed transmission had all the gear ratios mounted in front of the crownwheel and pinion, ahead of the rear axle line, as in contemporary F1 style, and a central gearshift was installed in the cockpit in order to provide a more precise change by eliminating the complexity of the linkage required.

The front suspension was activated by push-rods to torsion bars with the dampers mounted above the drivers' feet, with the major emphasis on ease of access for mechanics and engineers.

The XJR-14 was built around a carbon fibre composite monocoque structure manufactured by TWR's associate company ASTEC, this company at Draycott, near Derby, also producing the advanced composite nosebox, body panels and wings used by the 1991 Jaguar World Championship challenger.

However, one was inevitably bound to wonder what was the precise purpose of harnessing all this technology. Only eight FIA Sports Car World Championship races were held in 1991, and although four manufacturers took turns in the victory circle, it was an illusion to suggest that the category was healthy. Far from it, in fact, for the new generation of 3.5 litre contenders were to complete the gradual destruction of serious international sports car racing which had started some 20 years earlier.

Since the announcement of the 3.5 litre rules at the end of 1988, many doubts had been voiced about the ability of the series to attract a full field. With that in mind, into 1991, FISA extended the lift of the old turbo cars in an effort to bump up the size of the grid. The down side to this was the fact that another 100 kg was added to their minimum weight limit of 900 kg, although FISA later relented and allowed the old Porsches to run at a 950 kg limit everywhere except Le Mans, while keeping the 1,000 kg limit at all races for any other old-style cars.

The new 3.5 litre cars weighed only 750 kg, could use F1-style special brews instead of organizers' pump fuel, made quicker stops with gravity-fed refuelling, and were guaranteed the first 10 places on the grid. This all left the turbos no chance at all!

In another change, the so-called sprint races were trimmed in length from 480 to 430 km in order to ensure that the new 3.5 litre cars could make it through to the flag with just two fuel stops. And for the first time in three years, Le Mans was back in the Championship schedule – where TWR made a cheeky stab at taking pole position in a one-shot effort with the XJR-14, even though the main focus of their attention centred on the superbly reliable XJR-12s.

Jaguar successfully won the World Championship with the XJR-14 in 1991, but TWR withdrew from the contest the following year, leaving barely sufficient cars to run a shadow of a series. In fact, it was probably only because of assurances from a private team that they would be purchasing several of the ex-TWR cars that the series went ahead at all. But this particular deal fell through even before the first race, and no Jaguars appeared in the World Championship, although TWR continued to maintain a presence on the IMSA scene.

Mercedes's Trojan Horse

The arrival of Mercedes-Benz on the Group C scene during the 1980s was a gradual process, reflecting a degree of unobtrusive stealth on the part of the famous German motor manufacturer which had been traditionally reluctant to become seriously involved in motor racing ever since the golden years of 1954/55, which culminated in Pierre Levegh's 300SLR catapulting into the crowd and killing more than 80 spectators in the 1955 Le Mans race.

The conduit for the eventual return of the 'Silver Arrows' was provided by the Swiss motor racing engineer

Peter Sauber, who had been manufacturing sports cars ever since 1970. Having been involved primarily in the 2 litre category for a decade, in 1981 he expanded his involvement to design and build a special version of the 3.5 litre classic BMW M1 coupe, using a lightweight carbon fibre composite monocoque shell manufactured by the neighbouring Swiss company, Seger and Hoffman. In this car Hans Stuck Jr and Nelson Piquet won that year's Nurburgring 1,000 km race benefiting enormously from the fact that their mount was some 70 kg lighter than the equivalent standard model.

For 1982, Sauber moved into the Group C category with a 3.9 litre DLF-engined machine, but like other users of this Cosworth-made engine, the team suffered badly from vibration and all the ancillary problems that stemmed therefrom. The next few years were patchy, both from the viewpoint of participation and performance, but the turning point came at Le Mans in 1985 when Sauber arrived with the all-new C8 model which would represent a turning point in the history of his company's racing efforts.

Installed in the back of this machine was a 5 litre Mercedes-Benz twin-turbo V8, reputedly capable of producing in the region of 700 bhp. It seemed like an enterprising choice of power units, for although the V8s were maintained by Swiss-based engine specialist Heini Mader, there were rumours that the whole project had a degree of tacit support from the German car maker. The actual chassis design work was carried out by Leo Ress, who had worked with Sauber since 1981, and would remain in this key post within the company right through to their graduation to Grand Prix racing in 1993.

That maiden outing at Le Mans was not a success. Danish driver John Nielsen was at the wheel when the C8's undertray worked loose, causing the car to become airborne over the hump at the end of the Mulsanne straight. The Sauber executed a mid-air loop before crashing back on to its wheels, its chassis seriously damaged, but its driver unhurt. The following year Sauber expanded the operation to run a second car, but the C8 chassis had inadequate ground-effect qualities to enable it to compete with the Porsche and Jaguar opposition, although it generally proved extremely reliable.

In August 1986, Mike Thackwell and Henri Pescarolo notched up what was to be the first of many victories for the Sauber-Mercedes partnership, in streaming wet conditions at the new Nurburgring. It was a promising trailer

For the 1990 season the Sauber tag was dropped from the team's Group C programme and the superbly sleek Leo Ress-designed contender was dubbed the Mercedes-Benz C11. Based around a carbon fibre composite monocoque central cell, its wind-cheating profile was developed in a full-size wind tunnel with rolling road facility.

for the 1987 season, which was marked by the advent of the new Sauber-Mercedes C9, built around a much stiffer bonded aluminium monocoque and producing dramatically enhanced ground-effect properties.

The team failed to win a single race in 1987, but at least managed to come away from Le Mans with a new outright circuit record it its credit. At the end of the year, Sauber was dismayed to discover that it would be losing its sponsorship from Kouros, the men's toiletries brand owned by Yves St Laurent, but Mercedes would start to display more overt support of the sports car project for 1988 when the cars carried the colours of AEG, one of MB's associated companies.

This change in attitude had been prompted by the

arrival of Professor Dr Werner Niefer as MB's Deputy Chairman, and in addition to the substantial support through AEG, the technical back-up was enhanced with the engines being developed for enhanced economy using the Bosch Motronic 1.7 engine management system. Four factory technicians were assigned to the team, attending each race to tend the V8 engines, and MB also offered assistance in developing the Hewland VGC transmission, using new gears designed jointly by Mercedes and Staffs Silent Gears.

For 1990, Leo Ress produced what was in essence an evolutionary version of the C9, but also his first full carbon fibre composite design for the Sauber team. Having wrested the World Sports Championship from Jaguar in 1989, the new season saw the 'Sauber' nomenclature dropped. Although the organizational infrastructure remained unchanged, they were now official Mercedes-Benz entries for all the world to see.

Talking of the new Mercedes-Benz C11, Ress explained:

'With this car we had the opportunity to build a new machine for the end of the Group C formula, putting together all the experience of the last three seasons. Nobody else in the pit lane had this advantage. I cannot say that there is anything special about the C11, but everything fits nicely. It is a good package.

'Our first C11 prototype chassis was made in Switzerland, but we never raced it. It was too heavy, and it wasn't right because the rules changed. In January 1989 we decided to start again with a new monocoque, and we didn't begin our tests until the following winter. It didn't make any difference to our decision that it would only race for one season, because most of the money was already spent.

'We had to make sure that we learned everything we could about carbon fibre chassis technology, because we knew that we would need a completely new car for the 3.5 litre formula [which started in 1991]. We needed information about chassis rigidity, weight and so on.

'Maybe in 1990, it was easier to learn, because the pressure was not so great. With the CV9, which had an aluminium monocoque, we would have had some problems in 1990, especially with Jaguar. We could not have won so many races, I am sure. We were racing for Mercedes-Benz, so we had to do our very best. We could not take risks.

'The [carbon fibre composite] chassis of the C11 had double the stiffness of the C9, which meant that we had to be more careful with the selection of the springs and shock absorbers. We had to be more precise in the way we set up the car, but a stiffer chassis definitely offered us more in the way of advantage.'

The Mercedes-Benz flat-12 engine for the 3.5 litre Sauber C291 was a big disappointment and almost certainly contributed to the German car maker's withdrawal of front-line support for Sauber's F1 graduation in 1993.

As in 1989, the Sauber/Mercedes partnership swept the board during the 1990 season, the final year of the Group C regulations. The Swiss-German alliance won eight races and finished the year well advanced in the preparation of the new 3.5 litre Sauber C291 which displayed radical thinking in that it was powered by a totally new 180-degree V12 engine. As things transpired, this would come to be regarded as something of a technical abberation which never quite managed to deliver the goods.

Mercedes-Benz V8 engine development

Under the direction of Dr Hermann Hiereth, the technical director of the Mercedes-Benz racing programme, race development of the twin-turbo 5 litre V8 was monitored with increasing seriousness from the moment Peter Sauber first installed the engine into the C7 back in 1985.

Although it was to be another three years before Mercedes became officially involved in this Group C project, Dr Hiereth and his staff kept a paternal eye on the Sauber programme from the outset. In fact the first serious study of what Group C might have to offer the prestige motor giant came in 1984 when Peter Sauber approached MB with a request to use their wind tunnel to assess the aerodynamic performance of the Sauber C76, a car ironically equipped with a 3.5 litre BMW six-cylinder engine at the time. It didn't take Mercedes-Benz long accurately to assess the potential offered by such a programme. As Dr Hiereth explained to *Racecar Engineering* magazine:

> 'Group C made specified demands on engine performance which could be applied to passenger cars. We were already engaged in trying to find ways to maintain power outputs while using less fuel, and we realized straight away that a Group C engine project could be very useful to that research.'

Prior to Peter Sauber's involvement, Dr Hiereth's advanced engineering team had decided to assess the performance of MB's 16-valve type M117 engine (the 5-litre V8 which powered the top of the range limousines and coupes) when boosted by a twin-turbocharger configuration. Dr Hiereth:

> 'The first task was merely to see 700 bhp from the engine on the test bed, which was not difficult. However, we became deeply interested in the practical problems that were created. It was complex to make the engine stable while it was delivering power at this level, and particularly to reach the fuel consumption figures that would be needed for Group C racing. The project quickly became a real challenge to us.'

It did not take long for Mercedes to recognize that there were practical limitations to the amount of data that could be accumulated from work on the dynamometer. This coincided with requests from Peter Sauber for MB to release development engines for his new Sauber C8 project, and, appreciating the possible benefits, the Mercedes management authorized the programme to start. But it was made very clear from the outset that for the time being at least this would be a 'hands off' involvement.

Although lacking in previous racing experience, Dr Hiereth felt this scarcely handicapped the Mercedes-Benz Group C engine development programme. Although his department was having to use its engineering training to divine scientific answers to the problems, they had no preconceptions about the whole business, and as the MB team became more familiar with the technical problems, the more it fascinated them. Dr Hiereth admits:

> 'By 1987 we were making a fresh evaluation of the project. We went through the same process of study, with the four-valve version of the V8 available to us in the near future, to ask if we were happy with what we had. Had we wanted

to, we could have produced a totally new engine, but we concluded that we were perfectly satisfied with the V8 now that we had the four-valve heads available.'

Throughout the 1987 and '88 seasons the 16-valve type M117 version of the engine formed the basis of the programme, to be followed by the four-valve type M119 in time for the Suzuka race at the start of the 1989 season. Dr Hiereth explained:

'As soon as we were able to use the four-valve layout, it gave us an immediate improvement, perhaps two or three per cent over the previous engine. The only problem was that the additional camshafts and valves added about 20 kg to the overall weight. But it didn't make sense to look at exotic materials because we were looking for the best possible reliability. What was important was that the centre of gravity was much lower in the new engine, and that reduced the problem.'

In assessing the optimum engine within the Mercedes range to be used for the Group C programme, Dr Hiereth admits that consideration was given to the small, highly turbocharged, four-cylinder engine, or different variations of the V8 – either 4.2 or 5.6 litres – but, taking the fuel consumption element into consideration, they came back each time to the 5 litre as being the ideal choice. Dr Hiereth continued:

'Under race conditions, our V8 engines turned slowly, at 7,000 rpm, and up to 7,500 rpm in qualifying when fuel consumption, of course, was not a matter of importance. Normally for qualifying, we went from 1.5 to 1.8 to 1.9 bar boost, which gave us around 870 bhp, while in 1990 we raced with 700 bhp.'

The sophistication of the engine management systems employed in the Mercedes C11s progressed in leaps and bounds throughout the V8's career. In 1990 the new Bosch Motronic MP1.8 system developed jointly by Mercedes and the Stuttgart-based electronics specialists, featured no fewer than 23 microprocessors – instead of five – which improved the computing capability by about 50 per cent. Says Dr Hiereth:

'During 1989 we decided that we needed to carry out a further development of the knock-detecting system. We wanted to use a cylinder selective knock management, measuring and reacting to the pressure inside each combustion chamber, instead of using noise sensors or washers on the spark plugs, which are the methods employed in passenger vehicles. This involved separate ignition time for each cylinder, and there were problems simply because there was so much data.

'The system involves individual ignition coils for each cylinder, and therefore dispenses with ignition cables. It required a completely different wiring harness. As a result we could not use the C9 chassis to do this development work and had to wait for the C11 which was designed to have this harness.'

From the promotional and engineering viewpoint, it certainly did Mercedes-Benz no harm at all to be seen to be using what was essentially a production-based engine during the halcyon years of its Group C involvement with the Sauber team. Moreover, the 5 litre V8 displayed a remarkable level of reliability through that period, a pleasant contrast to the experiences of other manufacturers, in other racing categories, who have attempted to adapt production engines for racing purposes. As Dr Hiereth recalls:

'It was good to be able to say that we were using a production engine. We learned so much about the V8, although we couldn't transfer any knowledge of turbocharging because the production car unit is naturally aspirated. But we do know more about high loads on bearings, although we would have to design a new production engine to take advantage of this knowledge.

'By weight, our V8 racing engine was 60 to 70 per cent original, from production – the crankshaft, crankcase, cylinder heads – although they need special machining, and we add a special coating to the crankcase. What is important is that all the main parts were the same, or nearly the same as in production. What we changed, mainly, were the induction and exhaust systems, the oil system and the cooling system. We have been able to show our customers how powerful we could make our V8, how reliable in racing, and how economical.'

As a practical demonstration of Mercedes' engineering excellence, the use of that 5 litre V8 made singular good sense, but the advent of the 3.5 litre sports car regulations led Mercedes, just as it led Jaguar, into a technical exile at the hands of FISA's latest regulations.

The Mercedes board of management gave the all-clear for a new challenger for the 3.5 litre class as soon as FISA announced this proposed new class for 1991. That was as far back as 1988 and Dr Hiereth recalls that a total of 14 concepts were considered:

'We made a notebook (sic) of the parameters we wanted to achieve. The power output of the engine involved us setting a target of 650 bhp, combined in a package that would weigh in at the minimum 750 kg while enabling us to produce an ideal aerodynamic package.

'We considered eight, 10 and 12 cylinders, and finally reached the conclusion that the 180-degree 12-cylinder engine would offer us a low centre of gravity without any penalty concerning aerodynamic efficiency.'

The first such development engine started out as a V12 constructed from two experimental V6 engines which happened to be available in the company's research department.

Leo Ress concluded that a single, full-width venturi was more effective than two more conventional tunnels on either side of the car. He admits:

'It was hard work to get the package balanced, because the flat engine allowed a low centre of gravity, but the gearbox – the drive being taken from the centre of the crankshaft – makes it a little higher. We moved the gearbox a bit during initial tests in the wind tunnel, but when we looked at the result we redesigned the gearbox. In the end, however, we reached a good compromise.'

As far as chassis packaging was concerned, the C291 was yet another evolutionary version of what had gone before. Like most of its fellow Group C competitors, it was structurally an extremely strong car which produced excellent wind tunnel results and its engine, doubtful configuration or not, incorporated the very advanced TAG Electronics engine management system, future versions of which would be seen on the F1 McLaren-Ford MP4/8 from the start of the 1993 season.

Yet the C291 was not destined to go down in history as one of the world's most successful sports racing cars. At the end of the day, the Mercedes flat 12 engine was heavy, lacked power and was so horrifyingly complex to install in the chassis that, despite Leo Ress's professed ambition that it should be possible to replace in three hours, in fact it proved impossible to replace between practice sessions.

Nevertheless, in its final race, at the Japanese Autopolis circuit, Mercedes' two young lions, Michael Schumacher and Karl Wendlinger, gave the C291 the sole victory of its career. But by now, sports car racing had badly lost its way, so Sauber bowed off the Group C stage, taking a year off in 1992 in order to focus on developing a Formula 1 challenge for 1993. Mercedes toyed with the idea of a full-blown Formula 1 involvement, but eventually decided against it after taking into account the marketing problems facing its new range of S-class luxury saloons.

Mazda's rotary challenge

The Japanese Mazda company's rotary-engined machines introduced an unusual element of technical ingenuity on to the World Championship sports car stage, originally entering the fray after being encouraged by a class win at Daytona with their RX7 sports car as long ago as 1982.

At Le Mans the following year they appeared with their distinctive 7171C chassis powered by their 13B twin-rotor engine which was rated at 300 bhp. In 1984 the Mazda challenge was greatly expanded in conjunction with a tyre development programme with the American BF Goodrich concern, two Lola T616 sports cars being fitted with the rotary engines to run a mixed programme of US and European races. The latter were in the C2 class of Group C which had been introduced the previous year, ostensibly to encourage the less wealthy entrants, being allowed 330 litres of fuel per 1,000 km race.

In 1985, Mazda experimented with a fascinating twin-rotor, forced-induction version of their engine, using twin Hitachi HT20 turbochargers, which boosted the power of the 13B engine to around 550 bhp at 8,000 rpm. However, the complexities of attempting to run a turbocharged rotary engine within the constraints of a fuel consumption formula were considerable, and despite initial assertions from a Mazda director that a turbocharged programme would be implemented, this choice of technical route was subsequently reversed.

For 1985 the BFG tyre company moved into the top IMSA GTP class with Porsche 962s, leaving Mazda to run its own modest programme with the twin-rotor 737C models, yielding a 19th and 24th place finish at Le Mans.

However, for 1986 the programme went up a gear when Nigel Stroud, the British designer responsible for Richard Lloyd's composite chassis Porsche 956 and 962 specials, was commissioned to build a totally new car which would be powered by a three-rotor engine – each nominally of 1.3 litres – which would develop around 450 bhp. Stroud's knowledge of composite structures was put to good advantage and the new Mazda 757 was built around an aluminium honeycomb monocoque, with all-carbon fibre bodywork and many other components manufactured from weight-saving titanium.

This chassis endured in competitive action, albeit in progressively modified form, through to the 1988 season when Mazda raised its own engine development stakes even further by debuting an engine with four rotors, thus raising the nominal capacity to 5.2 litres and boosting the power output close to the 600 bhp mark.

Far from regular players on the Group C stages, it nevertheless fell to Mazda to produce one of the great surprises of the 1991 sports car championship season. Having mounted a token effort for the remainder of the schedule in order to guarantee their place at Le Mans, Mazda reaped the benefit of what was a very strategic investment indeed.

Having persuaded FISA to let the rotary-engined cars race at a low 830 kg limit, significantly less than the 1,000 kg cut-off point required for the Mercedes turbos and Jaguar V12s, the Japanese manufacturer pulled off a quite outstanding victory.

With F1 drivers Johnny Herbert and Bertrand Gachot being partnered by Volker Weidler, Japanese F3000 exponent, this trio never had a moment's trouble with the gutsy little 787B, which had been splendidly updated by Nigel Stroud. Although Mercedes set the pace from the start, the silver machines wilted with technical problems, leaving the Mazda to lead across the line the trio of Jaguar XJR12s, their huge 7.4 litre V12 engines unable to compensate for the legislated 170 kg handicap they had been saddled with.

Peugeot joins the fray

At the end of 1988, Peugeot decided to try its hand at the new 3.5 litre category and announced that a full programme in the World Sports Prototype Championship would be initiated from the start of 1991, preceded by a limited programme of events towards the end of 1990.

The heart of the new Peugeot 905 was a brand new 80 3.5 litre V10 engine, weighing in at less than 150 kg, developed by former Renault engineer Jean-Pierre Boudy. The twin overhead camshaft unit had a bore and stroke of 91 x 53.8 mm, intended to develop around 600 bhp from the outset. Technical Director Andre de Cortanze explained:

The Peugeot 905 was possibly the most sophisticated of all the 3.5 litre Group C spring cars, developed to a winning pitch for the 1992 season when it sped commandingly to the last Sports Car World Championship, its high-tech back-up ensuring that it comfortably out-performed its paper-thin opposition.

'We made a lot of investigations and calculations, and concluded for endurance racing purposes it was better to have a consistent, reliable 10-cylinder engine than a potentially unreliable 12-cylinder unit. It is also smaller, lighter and offers the possibility of much better fuel consumption.'

Designated SA35, the Peugeot engine was based on an aluminium monoblock – the block and crankcase being combined – with aluminium cylinder heads. Its five-pin, steel crankshaft ran in six plain bearings, with an aluminium sump forming each main bearing cap.

This power unit featured a Magneti Marelli engine management system using distributorless ignition with one coil per plug. Power was transmitted through a specially designed Peugeot transaxle, uniquely splitting the gearbox either side of the crownwheel and pinion first and second gears were ahead, third to sixth behind. Two longitudinal shafts are employed with the take-off pinion partway along the upper shaft and the crownwheel offset from centre. This layout was adopted to achieve optimum weight distribution and to position the CWP high enough to prevent problems with driveshaft angularity.

The transaxle casing was manufactured from magnesium and access to the forward gear cluster was facilitated by a detachable bell housing in which was incorporated the oil tank. Originally the drive went via a special Peugeot clutch pack/differential, activated by hydraulic pressure from an engine-driven pump which also served the power-assisted steering system, but this proved unreliable and excessively heavy, and a conventional ZF clutch pack/differential eventually replaced it.

The engine/gearbox package was installed in a carbon fibre/aluminium honeycomb chassis developed in close association with Dassault, the aerospace giant better known for its involvement with the Mirage fighter programme.

The body was largely the work of Peugeot chief stylist Gerard Weltier, previously known in motor racing for his WM Peugeot sports cars which had run at Le Mans since the early 1970s, and the car's aerodynamic profile was conducted with full and quarter-scale models in the wind tunnels at St Cyr and London's Imperial College respectively. Power steering was also quickly incorporated into the 905s racing specification, an unusual development.

The Jaguar XJR-14 was effectively a two-seater Grand Prix car, right down to using a Ford HB Grand Prix engine. It dominated the first year of the 3.5 litre sports car regulations in 1991, but the TWR team did not have the sponsorship or inclination to continue the battle against Peugeot into 1992.

When the 905 finally made its race debut towards the end of the 1990 season, in the hands of Keke Rosberg and Jean-Pierre Jabouille, many people were left wondering whether Peugeot had read the same rule book as their rivals, with liberal interpretation of such items as doors. The glorious wail of the new French V10 really did signal the start of a new era for the category and Rosberg, in particular, returning as he was from a three-year retirement, brought a breath of fresh air into what had increasingly become a rather insular series.

Up to this point, the perception of a 3.5 litre sports car had been based on the Cosworth DFL-engined Spices, but

over the winter of 1990/91 it became clear that Peugeot would have plenty of opposition from the new Cosworth-Ford HB V8-engined Jaguar XJR-14s. So it proved, with the TWR machines just pipping the new Peugeots to take the title.

Despite the fact that ex-McLaren F1 engineer Tim Wright was added to the Peugeot engineering staff over the winter, many people were surprised when the 905 appeared for the first race of the '91 season virtually unchanged from its preliminary outings the previous year.

The team's first full season started on a high note with a victory at Suzuka after the Jaguars wilted, but the fact that this was a fortunate success was rammed home at Monza where the 905s were well off the Jaguar pace. At Silverstone they would be thrashed once again and an outing at Le Mans, where Peugeot knew their machinery wouldn't last, at least gained them the consolation of leading in the early stages.

After the French classic, there was a two-month break before the next round of the World Championship, held at the new Nürburgring, and Peugeot certainly didn't waste their time. In a remarkable effort, the 905 was virtually transformed, taking a leaf out of the XJR-14 book from the aerodynamic viewpoint, and dramatically raising the level of their competitive performance.

These modifications were carried out under the direction of aerodynamicist Robert Choulet, the initial package having proved excessively sensitive to front ride height adjustments. The problems with the original aerodynamic configuration may have been compounded by the fact that initial development of the body profile was carried out on a fixed floor wind tunnel, but the Evolution 1 version, with its two-tier rear wing and secondary nose aerofoil, certainly made all the difference.

Peugeot competitions boss Jean Todt had promised early in the year that the team would make changes in time for the German race, but it is possible that even he did not expect them to be so outstandingly successful. With all-new bodywork copied shamelessly from the new Jaguar, the revised 905 at a stroke had more downforce and much enhanced handling.

That in turn enabled Michelin, the team's tyre supplier, to make better progress, and at Magny-Cours, Peugeot dramatically turned the tables on Jaguar and scored a crushing 1-2 success on their home soil. Suddenly, Peugeot had an outside chance of taking the title at Autopolis, but could finish only fourth after a troubled race.

Peugeot's prospects for 1992 looked strong, but, even before the end of the previous season, the very future of the Sports Car World Championship was seriously in doubt. By the end of the new season, even its staunchest supporters conceded that it was beyond resuscitation.

In November 1991, the newly elected FISA President Max Mosley, and Bernie Ecclestone, FISA Vice President of Public Affairs, had attempted to kill off the category, appreciating that it would be so thinly supported as to be almost valueless. Understandably, this caused a storm of protest, and at a meeting of the FISA Sports Car Commission, Mosley asked the assembled competitors whether they would be taking part the following year.

Tom Walkinshaw, unable to justify a budget from Jaguar and without a major sponsor, was not in a position to commit. Neither would Jochen Neerpasch, on behalf of Mercedes-Benz, who was then confident – wrongly, as things turned out – that the German car maker would back his proposals for a move into Formula 1. Peugeot's Jean Todt, meanwhile, had been thwarted in his attempt to get to the meeting by bad weather, and was trying to keep in touch with the proceedings by phone from Paris.

The end result of all this was that Mosley declared there was not enough interest and proposed that the SWC be cancelled, perhaps hoping that Peugeot could be persuaded that F1 was a more viable area on which to focus their efforts.

However, while Jaguar and Mercedes slipped quietly away, Jean Todt worked hard over the winter attempting to persuade the sports governing body that there would be sufficient cars. Toyota, Mazda and some privateers offered hope, but Alan Randall's plans to purchase a cache of TWR Jaguars came to nothing, and while FISA was eventually persuaded to change its mind, at the end of the day the high-tech Peugeot team found itself the lynchpin in a field of only 11 regular cars.

Whilst winning the Sports Car Championship against this makeweight opposition was quite a predictable outcome for the well-drilled Peugeot team, success at Le Mans proved to be the high spot of the year. It was only Peugeot's second outing at the Sarthe, so the success of their long-tailed 905 in this gruelling event must be regarded as all the more remarkable.

With all the World Championship sports car events now reduced to the level of inconsequential sprint races, Le Mans obviously placed an enormous potential strain on any 3.5 litre contender hoping to endure the 24-hour

The rotary-engined Mazda 767C won the 1991 Le Mans 24-hours, but the Japanese firm bought Jaguar XJR-14s for the following season, applied their own badges and went racing in the 3.5 litre category on what many people regarded as a rather tenuous basis.

marathon. Painstaking mechanical preparation has always been a cornerstone on which to build hopes for any realistic success in this event, so in terms of uprating their machinery, the challenge facing Peugeot at Le Mans was absolutely characteristic of the event.

To start with a strengthened engine was revised as the result of no fewer than six attempts on a 24-hour test run at Le Mans, only one of which saw a 905 last out the necessary distance to convince the team it had a realistic chance at the Sarthe. In addition, there was a heavier duty gearbox incorporated and dramatic revisions to the aerodynamic specification, the distinctive long-tailed configuration being finalized towards the end of 1991 before being introduced into the Paul Ricard test and development programme.

In order to out-run the opposition to take pole position, the Peugeots used qualifying fuel and sprint specification engines, thereafter reverting to their endurance engines to score a copybook victory. But sports car racing on this level had no future. Despite the fact that the category was so poorly supported, Peugeot and Toyota in particular, were hurling F1-type technology into the fray with such refinements as semi-automatic transmission (Peugeot) and traction control systems (Peugeot and Toyota), and set against the economy realities of international motor sports in the early 1990s, none of it made any real sense.

At the end of 1992 Peugeot withdrew from the sports car fray, sat back and considered its options. A Formula 1 programme was by this stage clearly on the cards, but the volatile nature of the Grand Prix technical regulations was certainly destined to be a matter of concern to the French company which, at the time of writing, has yet to finalize its future plans.

CHAPTER 12

FAREWELL TO TECHNOLOGY?

The Hilton Hotel at Terminal 4 of London's Heathrow airport may seem an unlikely venue in which to make motor racing history, but on Friday, 12 February 1993, the FISA Formula 1 Commission effectively sounded the death-knell for a decade of complex high technology development.

It was a decision which had been a long time coming. For years there had been many people – those committed to engineering ingenuity – who had marvelled at the way in which Formula 1 cars were becoming increasingly refined and complex. However, there was also a growing body of opinion which felt that the whole area of semi-automatic gearboxes, under-car aerodynamics and its reliance on the development of active suspension systems, drive-by-wire throttle control systems and traction control mechanisms were taking Grand Prix racing into an era where the racing was being jeopardized by the wide disparity of resources, and their consequent ability to pay for these developments, between the teams at the front and back of the starting grid.

Bestriding these two viewpoints, of course, was the Concorde Agreement, that set of rules and regulations which had governed the way in which Grand Prix racing was administered ever since the early 1980s. The Concorde Agreement had been hammered from the solid rock as part of a peace process uniting the sport's governing body FISA and the F1 Constructors' Association (FOCA) after two years of confrontation and disagreement over the way in which technical and sporting regulations should be interpreted between 1980 and '82. Its most fundamental cornerstone was a requirement that unanimous agreement was required to make instant rule changes, and that had functioned pretty well as a safety net throughout the decade that followed.

However, the catalyst for change came in November 1992, when Williams Grand Prix Engineering was a day late lodging its official entry for the 1993 FIA World Championship. Whether through sloppy office work or a misplaced sense of defiance, Frank Williams's organization had, at a stroke, put itself at the mercy of those of its rivals who viewed the team's domination of the 1992 World Championship with increasing concern. Remember, Nigel Mansell had used the superb high-tech Williams FW14B to sail to the World Championship, scoring nine wins out of 16 races.

Thanks largely to a stunning lack of imagination on the part of television directors at the individual Grands Prix, the racing was being portrayed as tediously processional to the countless viewing nations which took ready-made F1 lifeblood in an era where sport on television had become big, big business. Something had to be done, and if Williams was to get the unanimous agreement of its rivals to be accepted into the 1993 Championship as a late entry, then it would have to offer some concessions on the technical front.

Early in February 1993, Ferrari President Luca di Montezemolo reiterated his warning, that the famous Italian team could quit Formula 1 unless fundamental and wide-ranging changes were made to the regulations, to bring the sport back to what he described as its manufacturing roots. Speaking to a group of businessmen in Bologna, he said:

> 'There is nothing forcing us to remain in Formula 1 which must change its rules absolutely to return closer to the technology of mass-produced cars. Ferrari will never stop racing, but if things do not change quickly, we could also opt for other types of competition. We have reached a point where 95 per cent of the technology learnt from racing has no application to road cars.'

It was perhaps unfortunate that Montezemolo offered these thoughts only five days after Ferrari had emerged from a calamitous test programme at Estoril where the new type F93A, driven by Jean Alesi and Gerhard Berger, were five seconds a lap slower than Alain Prost's Williams.

Moreover, the apparent crisis enveloping Ferrari on the

track was mirrored in its industrial operations. The famous luxury sports car manufacturer had recently announced that it would have to lay off some 400 workers for the months of February and March 1993, this being the third time since November 1992 that Ferrari had been obliged to resort to such temporary lay-offs.

However, although the initial reaction to Montezemolo's remarks was one of cynicism, particularly within the British F1 fraternity, the truth was that the Ferrari President was only repeating views which had been made public on several occasions since taking over his job. And there were others within the F1 community who agreed with his assessment of the situation.

Perhaps understandably, taking all things into account, his observations met with scepticism from Frank Williams and his Technical Director Patrick Head, the latter challenging the notion that F1 necessarily had to have a direct link with road-going machinery.

However, taking a long-term view, Head revealed himself to be quite flexible in terms of what technical avenues might be adopted as a basis for long-term change in Formula 1. He conceded:

'I think a lot of people believe that we should look closely at the regulations, but this is a package which has to be discussed on a long-term basis. My view is that 1993 is already fixed, so let's start talking about 1994 or '95.

'We must reduce aerodynamic downforce by about 50 per cent in order to continue to ensure that this is a visual sport. We must give designers a brief to produce sensible-looking cars to sensible regulations, yet ones which open out the braking distances to ensure more overtaking.'

However, one man who wanted to accelerate the speed at which the winds of change should be allowed to blow through Formula 1 was Flavio Briatore, the boss of the Benetton Formula Grand Prix team. A recent recruit to the F1 game, Briatore was primarily a businessman with a businessman's feel for the way in which Grand Prix racing should evolve as a commercial operation through the 1990s. And he took a hard line. On the eve of that crucial F1 Commission meeting, he said:

'We do not want to see Williams out of the championship, but if that is the price we have to pay for some sensible changes to the Concorde Agreement, then so be it. If it comes to it, I would prefer to keep Ferrari and lose Williams.'

Things didn't look good for Frank.

What came next was quite remarkable and, with the benefit of historical perspective, may come to be regarded as Max Mosley's greatest day in his role as FISA President. Despite the fact that the meeting failed to reach a unanimous agreement, Mosley forced everybody involved to accept a degree of compromise in the interests of the sport's long-term good health.

The outcome of the meeting was a series of short-term cost-cutting restrictions on practice, number of tyres and limitations on the use of spare cars for 1993, but from the start of 1994 he came out with a huge list of changes, the effect of which was drastically to reduce the application of high-technology systems and swing the onus for performance back towards the driver in the cockpit.

The Formula 1 Commission voted by 10 votes – and three abstentions on the part of Frank Williams (Frank Williams), McLaren (Ron Dennis) and Bernie Ecclestone (FISA) – to introduce the following regulations from 1 January 1994:

---

1. No system or device may be fitted to the car which can control automatically any aspect of the car's operation or in any way mitigate the effects of an error by the driver except: (1) Systems concerned solely with the functioning of the engine. (2) Systems which only come into operation in the event of an accident which causes damage to the car. (3) Safety systems specifically approved by FISA.

2. No signal of any kind may pass between a moving car and anyone connected with the car's entrant or driver save for legible messages on a pit board or body movement by the driver.

3. The same engine must remain in the car for the entire event, subject to a maximum of 12 engine changes per season. The use of the spare car counts as an engine change. So does removing the pump, cylinder head or equivalent.

Further, from 1 January 1995:

The Formula 1 Technical Working Group will write a rule for a stepped undertray for Formula 1 cars. This will reduce cornering speeds, increase ride heights and make the cars less sensitive to ride height changes. The cars will then also be able to run on ovals without difficulty.

At a stroke, the death sentence, commuted for 12 months, had been imposed on semi-automatic gearboxes, active suspension, traction control systems and anti-lock braking systems. Those who reported that McLaren boss Ron Dennis emerged ashen-faced from this F1 Commission meeting only had to wait another 12 hours to see why. When the technical specification of the McLaren team's new Ford-Cosworth HB-engined MP/8 was published the following morning, it included not only all the aforementioned state of the art refinements, but also a secure, encrypted radio link allowing engineers to make changes to the cars' operating systems during the race. After 16 races, all these accessories would be consigned to the motor racing history books.

This was truly a turning point in contemporary Formula 1 history, a recognition that the sport had to relate more closely to outside pressures at a time when the industrial world was wrestling with the most severe economic recession of the post-war era. Mosley had effectively breached the Concorde Agreement in a successful effort to break a hardening stalemate from which there seemed no realistic possibility of escape.

In that sense it was less a peace treaty, more a suspension of hostilities – and one which had been imposed rather than reached by agreement and assent. Yet there would be more controversy, anguish and debate lurking further down the road as F1 wrestled to formulate a clear identity and purpose to take itself into the 21st century.

CHAPTER 13

WHERE NEXT FOR RACING SPORTS CAR TECHNOLOGY?

The death of the expensive 3.5 litre World Championship sports car category at the end of 1992 led FISA to think seriously about a viable alternative for the future. Plans for a European GT championship were offered, with the promise of a World Championship for such machines from 1995 if this proved a success. In response to FISA's proposals, Jaguar commissioned TWR to develop a competition version of what had been billed as the world's fastest production car, the Jaguar XJ220.

This car, in its final specification, explained *Racecar Engineering*, was planned to be eligible for the new GT championship, the Le Mans 24-hour classic, and various national GT series which were also tentatively planned. However, a close examination of the first such machine revealed this to be rather more than a lightly tuned high performance road car, much closer to a turbocharged Group C machine than many had expected.

The heart of the XJ220C was the 3.5 litre, twin-turbo V6 which had first been seen in the 1990 TWR Jaguar XJR14 coupe. The forerunner of the road-going XJ220 engine, this unit had the potential to produce around 1,000 bhp maximum, but in its Group C application was downgraded to develop around 750 bhp, depending on the requirements of the contemporary fuel formula. Detuned and detoxified for road-going use, the Jaguar coupe had a quoted output of 542 bhp at 7,000 rpm at a 2.0 bar boost pressure.

Right *The shape of things to come? The competition version of the striking Jaguar XJ220 supercar was revealed at the start of 1993 as a potential candidate for FISA's new International GT Championship. The idea behind this series is to bring sports car racing back to something the public can find more easily identifiable with road-going machines.*

For the XJ220C the engine was equipped with Zytek electronic, multi-point sequential fuel injection and engine management system and the water cooling needs of the twin Garrett turbochargers was satisfied by two side-

TWR

TWR

mounted aluminium air-to-air intercoolers.

For GT racing purposes, the Jaguar V6 output would be determined by the size of the relevant air restrictor, both the ACO (for Le Mans) and IMSA regulations calling for twin 32 mm restrictors while FISA, at the time of writing, had still to decree the restrictor limit for the European GT series.

The XJ220C was developed by TWR's Special Vehicles Operation group which took the basic silhouette and monocoque of the road car, the latter being an aluminium honeycomb structure owing more to the Jaguar team's racing pedigree than any specific road car technology.

This chassis incorporates machined aluminium bulkheads and an integral steel roll cage, conveniently conforming with FISA racing requirements.

In standard 'road-going' specification, the XJ220 carries its front suspension on the front of the monocoque, the rear attached to the engine subframe. For competition purposes, two rear end options were offered, one having a TWR Group C transaxle which was designed to carry the rear suspension. This bolts straight to the back of the engine and imbues the rear end with considerable torsional rigidity.

Ventilated and cross-drilled outboard-mounted Group C-derived steel disc brakes with radial-mounted six pot aluminium alloy AP Racing callipers were employed with aluminium alloy disc bells.

For Le Mans the ACO demanded a maximum wheel rim width of 12 in, but the FISA GT series regulations allowed the complete wheels to be a total of 24 in wide per side, IMSA 25 in per side.

While the XJ220 in road trim weighs in at 1,375 kg, stripped out for racing, with advanced composite materials replacing its original aluminium panels, the competition version comes down close to 1,000 kg the mandatory minimum for Group C at Le Mans in 1991. A 2.35 litre turbocharged GT car had to weigh in at 1,300 kg for the FISA series, but it was expected that the 220C would be able to run at a significantly lower all-up weight at both Le Mans and in the IMSA series.

The XJ220C also benefited from a full wind tunnel-derived underbody incorporating lightweight composite ground-effect venturi tunnels and a lightweight composite detachable nose and tail section for instant chassis access.

The degree of aerodynamic freedom included within the FISA and ACO regulations enabled TWR to modify the original XJ220 aerodynamic package for additional downforce, a slightly larger rear wing being used in addition to modifying the underwing, from splitter to diffuser. This aspect of the XJ220C development was carried out with a 40 per cent scale model in the Manchester University rolling road wind tunnel.

The Jaguar XJ220C was the first of this new breed of racing car to be unveiled, its relatively early debut hopefully encouraging rival manufacturers to throw their hat into the ring and support a class of racing which is designed to accommodate the public's perception of a high-performance, road-related supercar.

FISA's new rules had effectively taken international sports car racing back to the position it had enjoyed in the late 1950s and early '60s, where the technical lineage of the competing cars was directly traceable, a product which, although hardly run of the mill, could legitimately be described as a production car. Marrying this concept to the high-technology manufacturing processes available in the 1990s was inevitably something of a gamble, but the only feasible way in which international sports car racing could be guaranteed anything approaching a viable future.

BIBLIOGRAPHY

Blunsden, John *The Power to Win. The design, development and achievements of the Ford Cosworth DFV, DFX, DFL and DFY V8 racing engines* (Motor Racing Publications, 1983)
Cotton, Michael *Directory of World Sportscars Group C and IMSA Cars from 1982* (Aston Publications Ltd, 1988)
Hucho, Wolf-Heinrich, (Ed) *Aerodynamics of Road Vehicles* (Butterworth, 1987)
Huntington, Roger *The Design and Development of the Indy Car* (Fisher Publishing, 1981)
Nye, Doug *Cooper Cars* (Osprey Publishing, 1983)
Nye, Doug *McLaren. The Grand Prix, CanAm and Indy Cars* (Hazleton Publishing, 1984 and 1988)
Nye, Doug *Theme Lotus, 1956–86. From Chapman to Ducarouge* (Motor Racing Publications, 1978 and 1986)
O'Rourke, Brian P. *Designing for Survival* (Automotive Materials, 1991)
Robson, Graham *Cosworth. The Search for Power* (Patrick Stephens Limited, 1990)
Savage, Dr Gary *Composite materials in Formula 1 Racing* (Metals and Materials, 1991)
Spurring, Q. (Ed) *Year of the Silver Arrows. Sportscar World Championships 1990-91* (Q Editions Ltd, 1990)
Wells, Ken *Cosworth. Creative Power* (Prancing Tortoise Publications/Kewkar Racing, 1991)

PERIODICALS
Autosport
Chequered Flag
Racecar Engineering
Racer

INDEX